The Supply Chain Management Casebook

The Supply Chain Management Casebook

Comprehensive Coverage and Best Practices in SCM

Chuck Munson

Vice President, Publisher: Tim Moore
Associate Publisher and Director of Marketing: Amy Neidlinger
Executive Editor: Jeanne Glasser Levine
Operations Specialist: Jodi Kemper
Marketing Manager: Megan Graue
Cover Designer: Chuti Prasertsith
Managing Editor: Kristy Hart
Project Editor: Elaine Wiley
Copy Editor: Barbara Hacha
Proofreader: Sheri Cain, Anne Goebel
Indexer: Heather McNeill
Senior Compositor: Gloria Schurick
Manufacturing Buyer: Dan Uhrig

© 2013 by Chuck Munson
Publishing as FT Press
Upper Saddle River, New Jersey 07458

FT Press offers excellent discounts on this book when ordered in quantity for bulk purchases or special sales. For more information, please contact U.S. Corporate and Government Sales, 1-800-382-3419, corpsales@pearsontechgroup.com. For sales outside the U.S., please contact International Sales at international@pearsoned.com.

Company and product names mentioned herein are the trademarks or registered trademarks of their respective owners.

Printed in the United States of America

First Printing June 2013

ISBN-10: 0-13-336723-1
ISBN-13: 978-0-13-336723-2

Pearson Education LTD.
Pearson Education Australia PTY, Limited.
Pearson Education Singapore, Pte. Ltd.
Pearson Education Asia, Ltd.
Pearson Education Canada, Ltd.
Pearson Educación de Mexico, S.A. de C.V.
Pearson Education—Japan
Pearson Education Malaysia, Pte. Ltd.

Library of Congress Control Number: 2013935802

To my parents, Karl and Barbara Munson

Contents

             Retailer from Overcharging for Soda 156
             Chuck Munson, Washington State University

Chapter 4   **Short but Sweet** 161
Case 16   Ethical Product Sourcing in the Starbucks Coffee
             Supply Chain 163
             Dustin Smith, Washington State University

Case 17   Tmall, The Sky Cat: A Rocky Road Toward Bringing
             Buyers and Suppliers Together. 171
             Jianli Hu, Woodbury University
             Olivia Congbo Mao, Alibaba Group

Case 18   Make to Demand with 3-D Printing: The Next Big
             Thing in Inventory Management?. 180
             Tom McNamara, ESC-Rennes
             Erika Marsillac, Old Dominion University

Case 19   Airbus' Overstretched Supply Chain: Just How Far Can
             You Go Before Your Supply Chain Snaps? 184
             Erika Marsillac, Old Dominion University
             Tom McNamara, ESC-Rennes

Case 20   How to Keep Your Food Supply Chain Fresh 188
             Tom McNamara, ESC-Rennes
             Erika Marsillac, Old Dominion University

Case 21   The End of Lean?: Automobile Manufacturers Are
             Rethinking Some Supply Chain Basics. 192
             Erika Marsillac, Old Dominion University
             Tom McNamara, ESC-Rennes

Chapter 5   **Unique Challenges from Around the Globe** 197
Case 22   A Brazilian Dairy Cooperative: Transaction Cost
             Approach in a Supply Chain 199
             *Fernanda Pacheco Dohms, Pontifical Catholic University of
             Rio Grande do Sul*

             *Sergio Luiz Lessa de Gusmão, Pontifical Catholic University
             of Rio Grande do Sul*

Acknowledgments

I wish to thank Barry Render, Consulting Editor at FT Press, for encouraging me to tackle this project. Barry has been a mentor for me for longer than he realizes, and I greatly admire his lifetime of work helping to bring the fields of operations management and management science to the masses. I am also very grateful for my Executive Editor on this project, Jeanne Glasser Levine, for providing excellent guidance and suggestions while leaving me free to try to shape the contents of the book into my own vision. A huge thank you, of course, goes out to the 49 other contributors of the enclosed cases, without whom this book would not have been possible. It has been a true joy to meet (electronically) and work with so many wonderful scholars from around the world who all have been very responsive to my nagging requests. They are excited to share their work and insights with you, the reader. Finally, on a personal note, I am deeply indebted to my wife, Kim, for all of her encouragement and patience during this lengthy process that included many late nights. And I dedicate this book to my late parents, Karl and Barbara Munson. They always wanted to see me do something like this but didn't quite get to see the finished product in time. Anything positive that I've accomplished in life is due to them.

About the Author

Chuck Munson is a tenured Full Professor of Operations Management at Washington State University. His Ph.D. and MSBA in operations management, as well as his BSBA *summa cum laude* in finance, are from Washington University in St. Louis. He also worked for three years as a financial analyst for Contel Telephone Corporation. For two years, he served as Associate Dean for Graduate Programs in Business at Washington State.

Munson serves as a senior editor for *Production and Operations Management*, and he serves on the editorial review board of four other journals. He has published more than 20 articles in multiple journals, including *Production and Operations Management*, *Decision Sciences*, *Naval Research Logistics*, *IIE Transactions*, *European Journal of Operational Research*, *Journal of the Operational Research Society*, *Annals of Operations Research*, *European Journal of Information Systems*, *Interfaces*, *Business Horizons*, and *International Journal of Procurement Management*. His major awards include being a Founding Board Member of the Washington State University President's Teaching Academy (2004); winning the WSU College of Business Outstanding Service Award (2009 and 2013), Research Award (2004), and Teaching Award (2001); and being named the WSU MBA Professor of the Year (2000 and 2008).

Preface

Objectives of the Book

Over the past three decades, supply chain management has become firmly entrenched as a vital area of emphasis for companies. Many companies have risen to the top of their respective industries by forging effective supply chain management techniques into strategic weapons. In short, supply chain management means effectively handling the relationships between one's suppliers and buyers. But in practice, the field covers a wide range of issues, including supplier selection, purchasing, risk management, logistics, distribution, pricing, demand and supply management, and channel coordination.

Case studies remain a popular and effective means to study and analyze real business decisions. This book has been designed to provide a variety of interesting supply chain challenges. Taken as a whole, the 30 cases in the book touch on many of the important supply chain issues facing modern organizations. Individually, each case provides one or more self-contained challenges for management, leaving room for the reader to ponder the next best steps. Most of the cases are primarily qualitative in nature, while more than one-third of them have been specifically designed for quantitative analysis.

One of the most attractive features of the book is the truly global team of contributors. Twenty of our cases are written by authors currently residing outside the United States, including the countries and territories of Brazil, Canada, China, Ecuador, France, Germany, Hong Kong, India, Italy, Japan, South Korea, Spain, and Taiwan. Not only are many supply chains global in the first place, but by reading through these cases we see that managers around the world face many of the same challenges as everyone else.

This book has been designed to appeal to upper-division undergraduate or MBA-level courses in supply chain management or operations management. University instructors who adopt the book will have access to an accompanying set of electronic teaching notes for the cases, including suggested solutions for the quantitative elements.

We also expect that supply chain managers and business professionals in general will find the book to be of interest. The cases are full
of ideas for addressing sticky supply chain issues. And it always helps
to consider how to address challenges *before* being faced with them.

Although some are lengthy, many of the cases in this book are
intentionally designed to be relatively short and focused, allowing
the reader to delve directly into the issues at hand. This format also
facilities a wide menu of choices for instructors to assign combinations of short and long cases that suit their needs. All the cases have
been written as a basis for class discussion rather than to necessarily
imply either effective or ineffective handling of an administrative situation—that is for the reader to determine.

Organization of the Book

I have divided the 30 cases among five chapters. Many of the
cases could fit well in more than one chapter, but I hope that this
arrangement helps readers quickly locate cases that are most appeal-
ing to them. In addition, I have provided a table at the end of this
preface that identifies key characteristics of the cases so that readers
can search for cases under various topics.

Chapter 1, "Comprehensive Coverage of Supply Chain Issues,"
gets the book going with two cases that cover a multitude of supply chain issues. We start things off with the unique challenges of
the Salvation Army in Dallas, Texas. This case provides a thorough
understanding of the operations in this humanitarian supply chain.
To support its charitable activities, the nonprofit organization accepts
donations and sells them back to the public via thrift stores. The
unusual supply process is fascinating. Case 2 provides an in-depth
picture of numerous activities of one of the world's leading food companies, Perdue Farms.

Although every supply chain issue involves an inherent form of
risk, the cases in Chapter 2, "Supply Chain Risk Management," introduce some very serious risk management challenges. Case 3 describes
the unusual problem of collecting blood platelets from donors and
then getting them to the hospitals and (hopefully) into patients before
they expire less than one week later. The authors provide a large data

set of nearly 6,000 transactions that instructors can access via the electronic teaching notes. Case 4 provides a comprehensive picture of several risk management issues at Molson Coors. The case has numerous qualitative and quantitative elements to consider. Cases 5, 6, and 7 touch on a variety of risk management issues, with an emphasis on trying to match supply with demand at Toyota China, Cisco Systems, and the Italian fashion goods industry, respectively.

Chapter 3, "Supply Chain Analytics," contains the cases with a significant quantitative element. Case 8 applies queuing theory to address the problem of an electronic waste recycler being charged by the city for trucks that wait at its facility. Case 9 addresses an important issue that many of our textbooks avoid—how to alter inventory decisions in a multi-echelon (warehousing) environment. Case 10 addresses the optimal level of postponement under conditions of uncertain supply and demand. Case 11 applies the factor-rating method to a supplier selection problem. Case 12 provides an aggregate planning analysis. Case 13 applies cost analysis to help choose among three distribution scenarios. Case14 addresses a facility layout/redesign problem for a nonprofit organization in Ecuador. Finally, Case 15 explores the issue of double marginalization and how to implement effective channel pricing that will benefit all firms in the supply chain.

Chapter 4, "Short but Sweet," contains focused cases that cover several important issues. Case 16 provides a fascinating description of an ethical issue that more and more companies face—do they purchase in a situation where they know some unfair or difficult working conditions are occurring, even though the price is cheaper? Case 17 describes the ups and downs of an Internet supply chain exchange in the booming economy of China. Cases 18–21 can be thought of as a set. These concise gems can be read in the classroom, and each can lead to some great class discussions. Star Trek fans will appreciate Case18, as advancements in 3D printing may someday permanently change the way we approach inventory management. Case 19 focuses on Airbus and the inherent risks involved in creating a super-lean supply chain. Case 20 explores perishable inventory in the grocery industry. With similarities to Case 19 and with reference to recent disasters, Case 21 questions the once "sacrosanct" philosophy from the auto industry that any lean initiative is better—always.

We end the book with a little world tour in Chapter 5, "Unique Challenges from Around the Globe." Case 22 describes the workings of a cooperative of (primarily) small dairy producers in Brazil. Case 23 takes us to Ricoh in Japan with a focus on the importance of establishing and maintaining appropriate management processes for effective supply chain strategy implementation. Case 24 examines the supply chain risk management challenges at a mid-sized Italian manufacturer whose supply chain manager gathers ideas from the experiences of four other companies. Cases 25–28 are centered in India, focusing on supply chain strategy, city waste disposal challenges, attempts to convert suppliers to a cloud computing platform, and challenges in the spare parts supply chain of an auto industry manufacturer, respectively. Case 29 introduces us to the issue of expanding third-party logistics globally and the advantages that can accrue from forming a partnership with a foreign third-party logistics provider. Case 30 concludes the book with a description of the effects on a multinational logistics network after a roof collapses at a plant near Rome. The case also provides a theoretical framework with which to approach the general problem of dealing with such "macro risks."

Key Characteristics of the Cases

Case Number	Country/ Region	Supply Chain Risk	Ethical and Environ- mental Issues	Logistics Issues	Supplier Manage- ment Issues	Quantitative Methods
1	USA	X	X	X		
2	USA			X		
3	USA	X	X	X	X	Lead Time
4	Canada	X		X	X	Hedging
5	China	X		X		Inventory
6	Japan	X		X	X	
7	Italy	X		X	X	
8	Canada		X	X		Queuing
9	USA			X	X	Inventory
10	India	X			X	Inventory
11	USA				X	Factor Rating
12	India					Aggreg Planning
13	USA			X		Cost Analysis
14	Ecuador		X	X		Layout
15	USA					Pricing
16	USA		X		X	
17	China			X	X	
18	Ubiquitous		X		X	
19	Europe	X			X	
20	Ubiquitous	X	X	X	X	
21	Ubiquitous	X			X	
22	Brazil	X	X	X	X	
23	Japan	X			X	
24	Italy	X			X	
25	India					
26	India		X	X		
27	India				X	
28	India	X		X	X	
29	USA/ Taiwan			X		
30	Europe	X		X		

1
Comprehensive Coverage of Supply Chain Issues

Case 1

The Salvation Army in Dallas: The Supply Chain Challenges of a Non-Profit Organization[1]

Arunachalam Narayanan[†]

In October 2011, Major Carl Earp of the Salvation Army Adult Rehabilitation center was discussing the goals for the upcoming fiscal year with his advisory council. His goal was to generate enough revenue from the Salvation Army's family and thrift stores to pay for the religious mission at their facility and send about 31% of the revenue generated to their head office in Atlanta for strategic initiatives. The director of finance, Roderic Horton, was also present at the meeting. He provided the advisory council with the updated financial statements for the last two years and, at present, they could afford to send only 14% of the revenue to the head office. To add to this burden, in the previous month, they lost one of their stores to a freak accident, when a DART bus ran through the store after being hit by a wrong-way driver. Unfortunately, the driver of the wrong way vehicle was under insured and they lost a store which generated $40,000 in monthly sales. The Major and Roderic were looking to their advisors for some innovative ways to increase revenue or reduce their expenses to reach their goals.

---

[1] This case was developed solely for the purpose of classroom discussion. Some details of the case, including names of the organizations, have been disguised. This case is not intended to serve as endorsements, sources of academic or business data, or illustrations of effective or ineffective management of the personnel or organization.

[†] University of Houston, Houston, Texas, USA; anarayanan@bauer.uh.edu

Salvation Army: Origins and Purpose

The Salvation Army is a worldwide evangelical Christian church, whose doctrine follows the mainstream Christian belief and its articles of faith emphasize God's saving purpose. In 1865, the Salvation Army founders, William and Catherine Booth, both Methodists, believed that William was called by God to be an evangelist. They did not agree with the decision of Methodist officials that he should be confined to a local congregation, so they resigned from their Methodist ministry. Rev. William Booth was then invited to hold a series of evangelistic meetings in London and his services became an instant success. Soon, his renown as a religious leader spread throughout London. Initially, Booth had only 10 full-time workers but by 1874, the numbers had grown to 1,000 volunteers and 42 evangelists. They served under the name "The Christian Mission."[2] In 1878, the Mission's name was changed once more—to "The Salvation Army." The military name ignited members' imagination and enthusiasm, and uniforms were adopted and military terms were given to some aspects of worship, administration, and practice. This brought certain order and authority to their congregation. Thieves, prostitutes, gamblers, and drunkards were among Booth's first converts to Christianity.[3] His congregations were desperately poor and he preached hope and salvation to them. Even today, this is the main mission of the Army.

Today, the Army has a presence in 124 countries with more than 15,000 churches or corps, and is managed by the international head-quarters in London.[4] The international leader holds the rank of the General and is based at the headquarters in London. He or she is elected by a group of senior Salvation Army officers (called the High Council) for a term of five years or until their 68th birthday is reached after which he or she must retire. The General is assisted in policy making by an Advisory Council. To manage the governance of the Army, the international movement is divided into 50 territories. Each territory's command is led by a territorial commander. Territories

---

[2] About Salvation Army, http://www.salvationarmysouth.org/about.htm

[3] History of Salvation Army, http://www.salvationarmyusa.org/

[4] Statistics of Salvation Army, http://www.salvationarmy.org/ihq/www_sa.nsf

are then divided into divisions, with a divisional commander leading a team of administrative officers in each one. Each division encompasses a number of corps and other Salvation Army centers. Operations of The Army are supervised by trained commissioned officers. Officer candidates undergo two year training at one of the resident Salvation Army colleges which include theory and practice in field work. After candidates are commissioned their promotion is based on length of service, character, efficiency, capacity for increased responsibility, and devotion to duty. The ranks are lieutenant, captain, major, lieutenant colonel, colonel, and commissioner. Lay members who subscribe to the doctrines of The Salvation Army are called soldiers. Along with officers, they are known as Salvationists.[5]

In the United States, the origins of Salvation Army dates back to 1879, when Lieutenant Eliza Shirley left England to join her parents who had migrated to America earlier in search of work. She held the first meeting of The Salvation Army in America in Philadelphia in 1879. Soon, the Salvationists were received enthusiastically. In 1880, reports of the work in Philadelphia convinced Booth to send an official group to pioneer the work in the United States. Three years later, the official group was able to expand its operation into California, Connecticut, Indiana, Kentucky, Maryland, Massachusetts, Michigan, Missouri, New Jersey, New York, Ohio, and Pennsylvania. Today, the Salvation Army in United States is divided into four territories, Western, Southern, Eastern, and Central territory. The national commander and national chief secretary coordinate the activities of each territory, while each territory is governed by a territorial commander. As the general case, the territories are divided into divisions led by a divisional commander. In the United States, the operations in major metropolitan areas are administered by an area commander. Under his command, there may be a number of Salvation Army officers commanding corps, community centers, and social service centers.[6]

The corps community center is the local Salvation Army center of operations seen in most towns and cities across a country. Each week, a variety of people will meet at the corps community center for worship, cross-cultural services, fellowship, musical activities, and

[5] People of Salvation Army, http://www.uss.salvationarmy.org/uss/www_uss.nsf/

[6] History of Salvation Army, http://www.salvationarmyusa.org/

other events. There may also be a variety of community outreach activities, as well as character building activities for youth and adults. Emergency relief, emergency shelter, and other social services may also be available at the corps community center. Some of the large social services centers, senior citizens' housing, domestic violence and children's shelters, camps and rehabilitation centers are administered directly by divisional headquarters or by territorial headquarters.[7] The revenue for running these operations are raised through donations received from both individuals and corporations. In the U.S., most of these donations are in used items, which are in turn sold through their network of family and thrift stores. The Salvation Army has a history of free rehabilitation from alcohol and drug abuse. Thrift stores also provide the revenue to run these Adult Rehabilitation Centers known as ARCs.

United States Southern Territory and Salvation Army–Dallas ARC

The administrative center of the Southern Territory is in Atlanta, Georgia. The southern territory comprises 15 southern states and is divided into nine divisions. One of these nine divisions is Dallas, TX. In alignment with mission of Salvation Army (SA), the Dallas region's mission is mainly to support its ARC. The mission was first established in April 1907 as the Men's Industrial Home. This was basically a place for homeless men. Later, the Men's Social Service Center was opened in 1926. During this time, they began a therapeutic approach to substance abuse with the main component being a work therapy regime with a heavy religious or spiritual emphasis. They moved to the current location in 1976 and just a few years later, they changed the name to Adult Rehabilitation Center because they were making their services available to women. Today the region's mission is led by Major Carl Earp and he is ably supported by his lieutenant and SA employees. The ARC in Dallas supports about 136 inhabitants. Their rehabilitation program offers residential housing, work, and group and individual therapy, along with spiritual care. This prepares the participants to re-enter the society and return to gainful

---

[7] About Salvation Army, http://www.salvationarmysouth.org/about.htm

employment. Many of those who have been rehabilitated also reunite with their families and resume a normal life. To be admitted to the program, one has to have the desire to get help. They can be referred by families, friends, courts, clergy, and community leaders, or may simply walk into the facility. Every potential participant undergoes a comprehensive intake interview to ensure the ARC program is the best possible match for them. If they are not, they are referred to an appropriate program in the community. A long-term commitment of at least six months is required to rehabilitate and return to normal life.

To support their rehabilitation efforts in the Dallas region, the SA solicits donations from individuals, business, and organizations. Most of the donations are in terms of goods, which are then sold primarily at one of their seven stores (as shown in Exhibit 1-1). The proceeds from these sales are used to support their mission in Dallas.

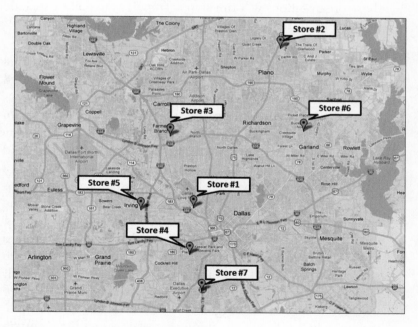

Exhibit 1-1 Stores of SA–Dallas ARC.

Donors

There are three types of donors: individuals, organizations, and businesses in the Dallas region. The list of items accepted at their facility is listed in their website.[8]

> *Individuals:* An individual can donate goods by either bringing them to the collection points of SA or by calling the collection warehouse to set up an appointment for pickup from their home or offices. The general donations include clothing, books, bric-à-brac, toys, furniture, beds, electronics, appliances, and sometimes even vehicles. They do donate food, but it is never resold in their stores.
>
> *Organization:* Several organizations around the city may donate their used furniture, appliances, and electronics. Sometimes these organizations will run some campaigns to collect specific seasonal items to donate to SA, like a Christmas toy drive.
>
> *Businesses:* They also donate used furniture and electronics to SA on a regular basis, but they are known for making occasional bulk donations. For example, a leading insurance firm donated a truckload of past football championship jerseys. These jerseys cannot be sold in the open market, as they have the wrong team names or they are way past its sales period. Another company donated a truckload of unique plastic flowers which it could not sell in its business. In 2008, Major Carl Earp established a significant relationship with a local transportation company[9] for their unclaimed items and lost baggage. The items left in the airline seat-back pockets and gate area like books, cell phones, mp3 players, CDs, jackets, and luggage that circle the carousel for hours without being claimed by the owners are kept at each airport facility for five days in hopes of being claimed by their owners. After that, the airline company sends the unclaimed items to their central baggage facility in Dallas. In 2008, the company began to give its items, not claimed within a certain period of time of arriving at the central baggage facility in Dallas, to SA in Dallas. Millions of coats, books, cell phones, gloves,

---

[8] http://dallas.satruck.org/donation-value-guide

[9] http://www.uss.salvationarmy.org/uss/www_uss_dallasac.nsf/vw-news/22007BD
3E2FB27568525744B005A9AA2?opendocument

and various other items are hauled off monthly by SA Employees and sold in local Dallas/Fort Worth SA facilities. In the last three years, this has turned out to be a sizeable revenue generator for SA–Dallas.

Not all donations are easily resold by SA, but they do not want to turn away these donors, as it may deter them from making future contributions. All these groups also donate monetarily, but that is usually less than 2–3% of SA–Dallas ARC's operating budget.

Reclamation Process

Every donation that arrives to the central warehouse in Dallas goes through a specific process of reclamation. Some are reclaimed and sold in the store immediately, while others have to be repaired before being sent to the stores. Several are in a bad shape and are sold in "as-is" condition at daily auctions, and the remaining are trashed at their facility. The following descriptions explain the reclamation process of the specific item types:

> *Reclamation of clothing items:* Clothing is by far the biggest donation in terms of number of items. They are also one of the top two revenue generators at their stores. SA stocks the clothing donations in a pile at their central warehouse facility and sort it manually. The employee visually inspects the clothes and sorts them as good, acceptable, and bad. The good clothes are in turn sorted in plastic drums (see Exhibit 1-2). Each plastic drum represents a certain dollar value, for example $1.99, $2.99, $5.99, $8.99, $10.99, or boutiques. Boutiques are high-end clothing like fur coats and wedding dresses and are sold at a reasonable price (up to $499.99) at their stores. The acceptable ones are later reamed and sold to exporters. Sorted reams are sold at $0.26/lb., while the unsorted reams are sold at $0.35/lb. In the sorted reams, there are no good clothes; SA has already taken the best ones to the store. The unsorted reams are priced a little high as there may be good clothing; SA didn't have the time to sort through all the clothing (because of high volume during holiday seasons). The remaining bad ones are generally sold in the daily "as-is" auction at the central warehouse; the leftovers after the auction are sent to trash. The entire process is depicted in Exhibit 1-2 and Exhibit 1-3.

2-A: *Salvation Army employee sorting the donated clothes.*

2-B: *Unsorted pile of clothes.*

2-C: *Ream of clothes.*

Exhibit 1-2   Clothing sort process.

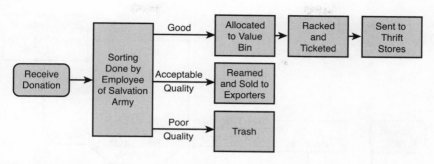

Exhibit 1-3 Reclamation process for clothing donations.

Reclamation of furniture and appliances: The other common types of items donated by all groups are furniture, electronics, and appliances. They are the highest revenue generator among the stores, but at the same time they occupy the largest floor space in both the stores and warehouses. Furniture, appliances, and electronics go through a similar reclamation process as depicted in Exhibit 1-4. A SA employee tests and inspects the product. If the product is in working condition it is sent to the store for sale in the next available shipment. If it can be repaired and if they have capable personnel in the warehouse, the product is refurbished and sent to the store. If not, it is sold along with the poor and obsolete inventory in the daily "as-is" auction at the central warehouse. The leftovers after the auction are sent to trash. Cell phones, music players, computer accessories all go through the same process.

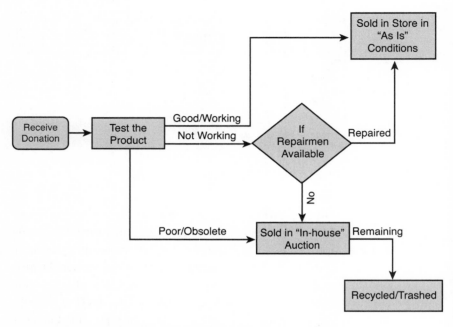

Exhibit 1-4 Reclamation process for electronics and appliances.

Reclamation of shoes: Shoes are sorted in the same way as clothing, but the only difference here is not all shoes are donated in pairs. The good pairs are sold in the stores, but a lot of the single shoes are bought out by third-party buyers in a separate auction.

Reclamation of books: Most of the donated books are classified as either paperback or hard cover. Paperback are sold at $0.99, while hardback are sold at $2.99 and up. One interesting aspect about the books is the ones donated by the airlines. They are usually the new bestsellers misplaced and unclaimed by the busy passengers at the airport. These are tagged with orange dots and stocked in the book racks of Store #1 and #2. They are generally the most desired books at these stores. With the advent of e-books, these opportunities may decline in future, but right now it doesn't seem to have affected the supply.

Reclamation of bric-à-brac: Bric-à-brac is collectibles and seasonal items such as vases, ornaments, and decorative objects usually displayed on the mantelpiece, on walls, or in cabinets. These are usually donated by individuals during a holiday or spring season when they clean their homes and offices. A seasoned SA employee inspects these items and selects the good ones for stores. The rest are sent to the daily auction site or trash. To attract customers to the stores during specific seasons, like Christmas and Easter, these items are sometimes stocked at the central warehouse for the right season to display at the store. For example, the Christmas items are usually displayed in the stores between early November and January. In this way, SA ensures maximum revenue from these donated items.

Stores

SA–Dallas ARC has seven family/thrift stores around the city of Dallas as shown in Exhibit 1-1, with four of them located in neighborhoods with a median household income of less than $43,000. These stores generate on an average $8 million in revenue every year, which are used by SA for its rehabilitation mission. The central warehouse is located near Store #1 in Harvey Hines Blvd and it is considered as the main store of operation for SA–Dallas. Almost all products are brought to the central warehouse for reclamation. Store #1 receives the best product, followed by Store #2, and so on. Inside the store, clothing displays receive first priority for placement, followed by furniture, bric-a-brac, shoes, and books. The revenue from each store by item category is provided in Table 1-1.

Table 1-1 Store Sales by Item Type in 2010 and 2011

12 Months Ending Sept 2010	Store #1	Store #2	Store #3	Store #4	Store #5	Store #6	Store #7
Clothing	$1,071,278	$411,358	$206,874	$164,568	$188,226	$260,768	$269,439
Furniture	$1,706,899	$617,097	$206,033	$180,671	$364,388	$283,375	$248,821
Bric-à-brac	$676,034	$234,690	$128,531	$84,881	$152,163	$140,662	$106,142
Shoes	$150,149	$81,305	$26,998	$24,666	$34,539	$27,741	$30,482
Books	$90,153	$23,196	$22,749	$7,753	$23,464	$17,673	$17,842
Total Sales	$3,694,512	$1,367,645	$591,185	$462,540	$762,781	$730,218	$672,725

12 Months Ending Sept 2011	Store #1	Store #2	Store #3	Store #4*	Store #5	Store #6	Store #7
Clothing	$1,131,303	$471,552	$217,133	$127,839	$198,427	$234,926	$264,509
Furniture	$1,572,563	$593,605	$210,330	$127,550	$368,858	$254,942	$255,609
Bric-à-brac	$727,253	$237,455	$142,940	$65,908	$162,249	$120,561	$116,089
Shoes	$148,830	$93,056	$29,523	$16,337	$36,799	$26,201	$33,224
Books	$79,728	$31,414	$24,547	$8,937	$25,241	$14,503	$17,326
Total Sales	$3,659,676	$1,427,081	$624,473	$346,571	$791,573	$651,134	$686,756

*Sales only until July 2011 (the store was destroyed in an accident)

To encourage inventory movement and customer traffic in the store, SA–Dallas follows a unique discounting strategy. Furniture is sold at a 25% discount after the product has been on floor for over 15 days, then at a 50% discount after 30 days and marked down to a 75% discount after 40 days or more. Very few furniture may remain after being marked off at 75%, those few items are ragged out and sold at their daily auction at the central warehouse. Clothing, electronics, bric-a-brac and shoes are color tagged to indicate the week of the month. After the first week, they are sold at 25% off; followed by 50% off after two weeks. After four weeks of being in the store, the product is taken (ragged) out of the store. If the ragged out item is still good, it may go to other stores; otherwise, it is sold at their daily auction. Therefore, one could safely assume the average inventory turn at these SA thrift stores is at least 12. This rapid change of inventory promotes repeat visit among shoppers who visit these thrift stores. Regular customers also know a lot about special deals at these stores, such as all clothing sold at 50% off every Wednesday or books from the lost luggage group being tagged with orange dots. Usually, these books are in pristine condition and generally fly off the shelves within a couple of weeks.

Other Sources of Revenue

Apart from the sales in the store, SA–Dallas also receives revenue from the donated goods in the following ways:

> *As-is auction:* At the central warehouse, every morning at 10 a.m. during the week days, the items that are not sent to stores are sold at an open auction in "as-is" condition. Usually, furniture, electronics, and appliances are the top pick in this auction. This type of auction generally raises about $60,000 in revenue every month. The items are sold at this type of auction mainly because there is no obvious market for them in the thrift store or the items are beyond repair. Sometimes items are also sent to this auction because there is no storage space in the warehouse or SA facilities. The items that are returned from the stores are sold in a separate auction either at the store or warehouse.

Junk and surplus: The items left after the open auction are pooled in a junk cart (Exhibit 1-5) and sold to the highest bidder. The buyers pay a price for every lb. (weight) of item in the cart. In the last few months, the team in SA–Dallas has started to recycle the metal (mainly aluminum) and cardboard in the junk. This new venture is providing them an additional source of $30,000 in revenue per month.

Exhibit 1-5 Junk cart being lined up for daily auction.

Secondhand vehicle sales: Some individuals also donate their old vehicles to SA. Every Thursday, SA runs a secondhand vehicle sale in "as-is" condition. This sale usually raises about $20,000 in revenue every month.

Specialty bulk item sales: Occasionally, big businesses around the city of Dallas donate items in bulk—for instance, truckloads of old sports championship T-shirts, plastic flowers, and misprinted company souvenirs. Many of these cannot be sold at their thrift stores and SA–Dallas has to find specific buyers for these products. Their retired colonel recently found an exporter to purchase 100 sports championship sports jackets at $4 each. They still have at least another 10,000 of these in their warehouse.

Challenges

The Major and his team in SA–Dallas ARC have established a sustainable revenue model to support their mission in their region. They support about 136 inhabitants at their center along with providing lawful employment for at least 188 people in the neighborhood through their warehouse, store, and rehabilitation center operations. All of this is done primarily using the revenue generated from the donated goods. Their mission is not without challenges, including the following they have faced in the last few years:

Managing peak periods: The donations from the individuals usually peak at the holiday season. This creates a huge bottleneck, both at their inbound and warehouse operations. They may not have enough vehicles to pick goods from the neighborhood, nor do they have enough labor to sort through all the donated goods. At these peak times, many goods will be sold in an "as-is" auction because of the lack of labor, space, and time to sort through the items.

Repair men: Lately, it's been difficult for the team to find repairmen to fix broken appliances and furniture (see Exhibit 1-6). Because of the lack of storage space, these items are sent to the daily auction place where it might fetch a lower value than what they could have earned by selling it at their thrift stores.

Exhibit 1-6 Appliances in the warehouse waiting for repair or to be sold in "as-is" auction.

Unsold items: SA–Dallas ARC has two warehouses, both located close to store #1. Both these warehouses still house a lot of unsold goods. These items have not been sent to trash because there is an obvious value to them; at the same time, the team is looking for new avenues to sell them at places other than their thrift stores. Some examples of valuable surplus items include electronics like computer printers, monitors, copiers, and cell phones. These items are not sorted at the thrift stores and are sold mostly in the "as-is" auction site at their central facility.

Cost of dump fees: The central warehouse disposes about 700,000 lbs. of trash every month and they pay an average of $21 per ton for dumping it. Around 20% of donations are sent to trash and another 10% are recycled. The remaining are either sent to store or sold at their daily auction in "as-is"/junk status. As the amount of trash increased in the last few years, the central warehouse spent $160,000 to build a trash compactor in-house to reduce the volume of trash before loading on to the dump trucks. All these infrastructure and monthly payments are made from the revenue generated from the business model without any subsidies.

Store #4 accident: In July, they lost a store to an unfortunate incident with a DART bus and an uninsured wrong way driver. Luckily, it was early in the morning, so no employees or customers were there at the time of the incident. The store was severely damaged (see Exhibit 1-7), and it was not economically viable for SA to repair and insure the shop for future use. They not only lost a store with $40,000 in monthly sales, but also lost a storage/display place for their items. The items that would have been sold at Store #4 would either lay idle in the warehouse or would get sold at the "as-is" auction site because of lack of storage space in their operation.

Exhibit 1-7 Salvation Army store damaged by an accident in July 2011.

Joel Prince/*The Dallas Morning News*—Reprinted with permission.

Soliciting needed items: The SA team in Dallas generally knows from experience the kind of items that gets sold in their stores during every season. But, they have very little control on the timing of the donations of these items. They do their best to stock these items and display them at the right season subject to their inventory flow and space constraints. They are still looking at innovative ways to solicit the items at the time of their needs. Recently, they have run campaigns in local newspapers soliciting car donations and general donations at their stores.

Surplus target: Every year in the U.S., SA southern territory headquarters provide a surplus target for its member regions. This year, they want the regional missions to generate enough revenue to support their local activities as well as send about 31% of the revenue back to the headquarters. This a voluntary target, and every mission strives to do its best. Last year, SA–Dallas ARC managed to send about 17% of their revenue to their office in Atlanta, Georgia.

Advisory Council Meeting in October 2011

The advisory council of the SA's Dallas ARC is composed of volunteers from the business community in the city—usually presidents, CEOs, and senior level managers. They meet on the third Tuesday of every month, from September to May, at their main ARC facility in Dallas to provide feedback and advise on the initiatives taken by the Dallas ARC. During this meeting, Roderic would present the financial statements, store and general revenue summary of Dallas ARC for the last two years (shown in Tables 1-2 to 1-3). The meeting held in 2011 was all about increasing revenue; several ideas were discussed, and both Major Carl and director of finance Roderic made a note of all the ideas. At the end of the meeting, Major and Roderic looked at their notes (see Exhibit 1-8) and were wondering which ones to target first and how to move forward.

Exhibit 1-8 Advisory Council Meeting Notes

Processes to Improve	Expenses	Revenue	New Opportunities to Look At
Recycling Reclamation process of regular donated items like clothing, books and bric-à-brac	Dump expenses Hiring professionals to repair or fix furniture and appliances	Opening a new store Salvaging the specialty donated items stocked in their warehouses	Use of social networks for soliciting donations Establishing new corporation models like the existing one with the public transportation company Selling unsold electronics/appliances and other collectibles through other means Establishing relationships with business entities for recycling like Goodwill's partnership with Dell[10]

[10] Dell reconnect, http://reconnectpartnership.com/

Table 1-2 Revenue of the Dallas ARC in 2010 and 2011

	Oct'09–Sep'10	Oct '10–Sep'11
From Stores	$8,281,606	$8,187,265
As-is Auction	$727,029	$640,383
Rags	$350,669	$520,359
Surplus	$39,623	$54,440
Junk	$56,374	$204,237
Corps Welfare Vouchers	$12,340	$14,149
Donated Vehicles	$272,987	$228,910
Other	$0	$4049°
Total revenue	**$9,740,652**	**$9,853,809**

*Proceeds from selling corporate specialty items/donations (past sports championship T-shirts, jackets, plastic flowers, and so on)

Table 1-3a Salvation Army—Dallas ARC Financial Statements: Assets and Liabilities

Assets	Ending Sept. 30, 2010	Ending Sept. 30, 2011
Bank Balance - Operating Account	$197,057	$155,634
Bank Balance - Payroll Account	$1,200	$1,200
Bank Balance - Petty Cash	$9,000	$5,000
Individual Petty Cash	$200	$200
Temporary Cash Advances	$200	$0
Store Change Fund	$5,290	$5,290
Accounts Receivable Local	$2,516	($4,634)
Accounts Receivable SA Other	$6,318	$10,526
Accounts Receivable THQ	$14,349	$107,463
Food Stamp Inventory	$2,334	$203
Total Assets	**$238,463**	**$280,883**

Liabilities	Ending Sep 30, 2010	Ending Sep 30, 2011
Accounts Payable Local	$91,952	$163,031
Accounts Payable Non-Automated	$4,323	($4,981)
Accounts Payable HQ Transmission	$30,283	$15,147
Accounts Payable SA Other	$0	$0
Accounts Payable THQ	$19,961	$12,084
Accounts Payable Unclaimed Payroll	$731	$385
F.I.C.A.	$0	($67)
State And Local Sales Tax	$52,227	$54,726
Employee Medical Insurance	$1,203	$2,431
Accrued Expenses	$37,784	$38,126
Total Liabilities	**$238,463**	**$280,883**

Table 1-3b Salvation Army—Dallas ARC Financial Statements: Income Statement

Income	Oct'09-Sep'10	Oct'10-Sep'11
Donations - General	$20,337	$17,197
Meeting Collections	$12,752	$11,490
Child Sponsorship Income	$6,889	$7,003
Grant - ARCC	$2,850	$10,185
Grant - Area/City Command	$7,535	$0
Unrestricted Trust Funds	$8,962	$6,489
Food Stamps	$215,878	$207,142
Fees - Meals	$126,055	$96,626
Vending Machine - Canteen Sales	$56,111	$53,550
Sales - Stores	$8,281,606	$8,187,265
Sales – As-Is Auction	$727,029	$640,383
Sales - Rags	$350,669	$520,359
Sales - Surplus	$39,623	$54,440
Sales - Junk - Center	$56,374	$204,237
Sales - Donated Vehicles	$272,987	$228,910
Sales - Corps Welfare Vouchers	$12,340	$14,149

Income	Oct'09-Sep'10	Oct'10-Sep'11
Sales - Other	$0	$4,049
Miscellaneous	$11,483	$2,019
Total Income	**$10,209,478**	**$10,265,492**

Table 1-3c Salvation Army—Dallas ARC Financial Statements: Expense
Statement

Expense	Oct'09–Sep'10	Oct'10–Sep'11
Officer Allowance	$133,661	$140,129
Employee Salaries	$3,079,158	$3,256,614
Pension and Retirement	$111,029	$128,988
Employee Insurance	$154,508	$167,029
Medical and Hospital	$694,564	$783,973
F.I.C.A and Other Related Taxes	$217,967	$229,684
Professional, Legal and Audit Fees	$131,076	$140,030
Data Processing Fees - Ceridian	$40,431	$42,377
Uniforms	$35,512	$43,692
Food and Beverages	$442,594	$464,778
Goods Purchased for Resale	$30,081	$34,349
Office Supplies	$238,982	$250,031
Kitchen Supplies	$31,624	$41,758
Miscellaneous supplies	$97,994	$99,454
Telephones	$83,463	$85,796
Postage, Messenger and Delivery Services	$12,260	$13,647
Building Rental and Insurance	$578,573	$595,211
Utilities	$575,088	$568,101
Property Upkeep, Repairs	$256,184	$256,079
Real Estate Taxes	$20,700	$11,991
Janitorial Supplies	$50,470	$70,127
Dump Fees	$104,033	$107,597
Rentals of Furn/Fix & Equip	$30,060	$29,996
Repairs/Maint. Furn & Equip	$22,384	$11,898
Purchases - Furn & Equip	$73,357	$71,466
Printed Materials and Subscriptions	$5,336	$7,170

Expense	Oct'09–Sep'10	Oct'10–Sep'11
Printed Material-Auto Ads -Newspaper	$70,657	$76,949
Vehicle Operating Cost & Insurance	$522,088	$543,573
Councils, Conferences, & Institutes	$25,408	$30,359
Assistance to Clientele	$156,082	$170,285
Scholarship Grants/Tuition	$4,200	$7,550
Christmas Remembrances	$8,230	$9,075
World Service Expense	$42,408	$44,412
Bad Debt Expense	$571	$2,643
Grants - ARCC	$4,800	$0
Vehicle Depreciation	$229,935	$277,475
Total	**$8,315,468**	**$8,814,286**
Surplus to Headquarters	$1,894,010	$1,451,206
Percent of Income	**18.55%**	**14.14%**

Discussion Questions

1. What are some of the channels you would choose to boost the revenue of Salvation Army's Dallas ARC? Which ones would you target first and why?

2. Comment on the donation, reclamation, and selling process of the different items. Are there any ch anges you would recommend to improve the effectiveness of reclamation? If so, explain them.

3. In your opinion, which of the current operations is the most efficient, and which one is the most expensive?

Acknowledgments

We thank Major Carl Earp, Roderic Horton, and his team for their support during the development of this case. We also thank Brent Burns, member of the Dallas ARC advisory council, for setting up this collaboration.

Case 2

Perdue Farms: A Vertically Integrated Supply Chain

Ling Li[†]

Products and Markets

PERDUE® is one of the quintessential icons in the American food industry. As one of the largest private companies in the United States,[1] Perdue Farms is a leading international food and agriculture business. It has 14 food-processing facilities in 12 states, employs more than 20,000 associates, partners with 7,500 independent farm families, and produces about 2.7 billion pounds of chicken and turkey annually. Its operating subsidiaries provide quality products and services to retail, food service, and agricultural customers. Perdue Farms supplies chicken products in more than 40 countries to chain restaurants, national and regional foodservice distributors, institutions, and the travel industry.[2]

Perdue Farms has processing, further-processing, and cooking plants throughout the eastern half of the United States. Though best known as a poultry company, it is a major producer of agricultural products as well. Perdue purchases grain from more than 5,000 local farmers to supply its feed mills and market grain internationally. It also owns a fleet of barges, leased rail cars, and a deep-water port to export grain and agricultural products around the world.

[†]   Old Dominion University, Norfolk, Virginia, USA; LLi@odu.edu

[1]   http://www.forbes.com/lists/2006/21/biz_06privates_Perdue-Farms_
       W8L7.html

[2]   Interviews conducted during a visit to Perdue Farms in 2005.

Markets

Perdue serves customers domestically and internationally. It exports food products to more than 40 countries around the world and creates products that meet local preferences, including shipping Perdue products to U.S. military installations overseas. Perdue's International Unit listens to customers and adapts ways of doing business to accommodate their special needs. Such needs include dual-language packaging, localized seasoning and breading preferences, certified processing to meet religious and local government standards, and portions prepared using local weights and measures.

Perdue has been a pioneer in many fields of the food industry. It has been the first to successfully brand and advertise select commodity products, the first to have the company leader become a celebrity advertising spokesperson, the first American company to produce retail fresh tray-pack chicken using a Chinese breed of chicken in China, and the first American company to market poultry products in Chile.

Food Products

Perdue's chicken products include fresh, marinated, frozen, fully prepared, and delicatessen chicken products. Perdue chickens are the meatiest chickens in the foodservice industry with the highest meat-to-bone ratio. By going from the freezer to the plate in minutes, foodservice operators are saving time and money on America's favorite appetizer or light entrée.

Through 80 years in the poultry business, Perdue has developed a stunning range of product offerings for the foodservice industry. Perdue Fresh Chicken has set the standard for freshness and quality for years. Further processed chicken products include roasted chicken, chicken fillets, breaded chicken, chicken wings, and more. Perdue upholds the values of product variety, consistency, and quality. Examples of its new further processed products include such items as Short Cuts Grilled Chicken Breast Filets, Char Grilled; Short Cuts Grilled Chicken Breast Filets, Honey Roasted; and Perdue Wingsters Split Chicken Wings, Buffalo Style. These fulfilling entrees are ready in only 8 to 12 minutes.

Agricultural Products

Perdue's Grain & Oilseed Division buys domestic grain from local farmers and dealers for further processing, feed manufacturing and domestic and international merchandising. It produces feed-grade soybean meals and crude soy oils, and it refines edible vegetable oils and produces lecithin for food companies. Perdue also operates protein conversion plants that manufacture livestock and pet-food ingredients.

Venture Milling, a wholly owned subsidiary, blends protein-based feed ingredients for livestock industries. Products include Pro-VAAI, a bypass protein for the dairy industry. Perdue-AgriRecycle, a joint venture, converts surplus poultry litter into organic pelletized and granular fertilizer products for agricultural and horticultural customers. Heritage Breeders, LLC, a wholly owned subsidiary, sells Perdue broiler breeder stock to other poultry companies.

New Product Research and Development

Research and development is conducted at the Perdue Innovation Center. Through research and innovation, new food and agricultural products are created while setting the highest standards for quality, food safety, environmental stewardship, and poultry welfare.

Perdue has spent decades perfecting ways to add value to its chicken products, including making it easier for customers to prepare the products for the end consumer.

Vertically Integrated Supply Chain

Raw Materials: Eggs

Perdue is a vertically integrated company. It owns the production from the egg to the finished products. The entire supply chain starts from breeding eggs, hatching chicks, manufacturing processing, packaging, warehousing, and distribution. This gives the company total control of quality. It has its own breed of chicken to supply eggs to hatcheries.

Chickens are raised in temperature-controlled houses on family-owned farms by approximately 2,500 family farms that benefit from their relationship with the company. Birds are raised in houses that are designed to keep the birds as comfortable as possible. They are automatically fed with feed specially formulated for their age, and nipple drinkers dispense water with a push of a button. The birds reach processing weight in four to six weeks, depending on the needs of the processing plant in the complex.

Poultry feeds are manufactured in house, and they are carefully formulated to match the nutritional needs of chickens at every growth stage. The flock supervisors are backed by an advanced team of scientists and laboratory technicians working with the industry's leading research and analytical equipment. Raising healthy birds and ensuring adherence to strict standards for food safety, poultry welfare and environmental stewardship comprise Perdue's commitment to quality.

Perdue maintains quality control through its distribution network, including its own truck fleet, distribution and replenishment centers and dedicated cold storage and export facilities.

Manufacturing Processing

The manufacturing process includes two separate production stages. The first hatches chicks, and the second kills, cleans and packages the birds. The killing-cleaning process includes receiving and killing operations, inspection, removing the internal organs of the poultry, cutting and deboning, and packaging. Many of these operations are automated, while others are done manually. The manufacturing process is described in the next section.

Packaging

Packaging is the next step to get the processed poultry from Perdue processing plants to consumers. It is generally a two-part procedure. First, the bird or bird parts are placed in a bag or package. Second, the package is placed in a shipping box. Poultry is packaged in a wide variety of formats, depending on customer orders.

In general, there are four types of packaging: bulk packaging, individual packaging, bone-in products, and bone-out products. For the whole bird bulk packaging, birds are bulk boxed and sent to large users such as Sam's Club or secondary processors. For the whole bird individual packaging, the bird is individually bagged and boxed for supermarket sale. Bone-in product are packaged and sold as consumer product or as bulk sale for large commercial users, as well as bone-out products. No matter how a bird is packaged, it is almost always placed in a large cardboard box for shipping. The general packaging steps include filling, weighing, and stacking these boxes.

There are a few "firsts" in the packaging process for which Perdue is proud. The company was the first to conduct research on a leak-resistant package, the first to package fully cooked chicken foods in microwaveable trays, and the first to use modified atmosphere packaging to ensure freshness.

Warehousing and Distribution

After the chicken is packed in its shipping container at the processing facility, it is moved from the processing floor to the storage area. If immediate shipment or placement in a warehouse is needed, the packed chicken is moved to a truck right away. It can also be shipped to Perdue distribution centers for storage until a customer needs it.

Perdue Farms has four distribution centers across the United States. The company equips its fleet of trucks with satellite tracking systems to ensure on-time delivery. A replenishment center is set up for retail customers. Perdue uses technology to communicate with customers, staying in touch via telephone, e-mail and video conferencing. Some stores have vendor-managed inventory control systems, which allow Perdue to track sales in real time.

The Manufacturing Process

Hatchery

On Perdue breeder farms, strict rules and precautions are followed to prevent the spread of diseases from other flocks. The eggs are collected and then sent to a Perdue hatchery. The hatchery usually hatches chicks four days per week, leaving one day for maintenance. Quality assurance starts at this point of the production process. Eggs are color coded with an account number. In case there are any quality-related issues, the problem can be traced to the root.

At the hatchery, eggs are put into incubators to start the growing process. For perfect hatching conditions, the temperature is set at 99.5°F and humidity is controlled at 86 percent. In the hatchery, the chicks are inoculated against disease right in the egg and are treated to prevent respiratory problems. The hatching process takes 21 days. After that, the chicks are carefully placed into Perdue trucks for their trips to family farms to be grown to processing weight.

At the family farms, when the birds reach processing weight in four to six weeks, depending on the needs of the processing plant, they are shipped to manufacturing processing facilities.

Processing

Receiving and Killing

The receiving and killing operation is largely automated, including receiving live birds, killing, defeathering, and removing feet. This operation includes the following tasks: (1) move the poultry cages from trucks to dumping areas; (2) unload live birds from the shipping cages to the conveyors; (3) manually take live birds from conveyors and hang them in shackles; (4) automatically kill poultry and manually kill any birds missed by the machine; and (5) automatically defeather the birds. Once the birds are defeathered, they are moved to the evisceration processing section via conveyor belt.

Evisceration

Evisceration processes remove the internal organs of the poultry. Hearts, livers, gizzards, and necks may also be cleaned and packaged in evisceration. This operation includes the following tasks: (1) automatically cutting the bird open; (2) cutting the neck; (3) cutting and removing the oil sack from the birds; (4) removing the viscera from the body cavity and arranging them for USDA inspection—every third person that inspects the viscera in the production line is from the USDA; (5) separating giblets and viscera; (6) removing and cleaning the gizzard; (7) washing and visually inspecting hearts and livers before they are sent to the bagging station; (8) placing hearts, livers, gizzards and paws into bags; (9) removing the lungs and the kidneys from the body cavity using a suction device; and (10) verifying that the carcass is eviscerated.

Cutting and Deboning

After a chicken has been eviscerated and cleaned, it is prepared for packaging as a whole bird, or bone-in products, or bone-out products. If the final products are bone-in or bone-out products, this involves one more step: cutting; or two more steps: cutting and deboning. (i) *Cutting:* In the cutting process, the wings, legs, and thighs are removed from the carcass and the back is cut away from the breast. At this point, parts such as wings or thighs can be packaged as a consumer product, bulk-packed for delivery to other processors, or shipped to other parts of the plant for further processing. (ii) *Deboning:* The deboning process involves cutting meat away from the bone and cleaning. The deboned parts such as the chicken breast are generally packaged as a fresh or flash frozen consumer product.

Food Safety and Quality Control

As a vertically integrated company, Perdue Farms is able to ensure quality at every step in the supply chain. Perdue Farms, Inc. is the only major poultry company with a unique proprietary breed of chicken. This ensures quality and consistency of the product. It is also the first poultry company to successfully implement a comprehensive "farm-to-fork" food safety program. The company has created a Food Safety Pledge to make sure that every associate has ownership of product quality and food safety.

At Perdue Farms, quality assurance managers are trained and are certified in Food Safety. In addition, each Perdue facility has a food protection manager, certified by ServSafe, the leading food safety education program. Quality and safety are stressed under the watchful eye of USDA inspectors. Each processing and further-processing facility has onsite USDA inspectors.

The quality program at Perdue Farms is meticulous at every phase. For example, the quality of products is ensured by inspections conducted by the company's product lab, digital scales are used to assure and guarantee accurate box weights, satisfaction of products is guaranteed through providing consumers a money-back option and an accessible toll-free consumer hotline, and nutritional information is labeled on every package.

Every day, a clean-up crew comes into food process sites to wash every nook and cranny. After cleanup is complete, plant laboratory personnel check to see that the job has been done properly.

Creating an environmentally friendly alternative use for surplus poultry litter is an example of Perdue Farms' commitment to environmental stewardship. The company invested $12 million to build a plant that converts surplus litter into organic, environmentally friendly fertilizer products for application where nutrients are needed.

Investment in Collaborative Planning, Forecasting, and Replenishment (CPFR)

Before investing in CPFR software, Perdue's managers determined their production volumes by the "gut feel" of Perdue's suppliers and customers, as well as the seasonal history of past consumption. It worked. However, with increasing competition in the food industry and higher demand from customers, Perdue seriously considers investment options in technology.

Since the late 1990s, Perdue's CIO, Don Taylor, has led Perdue's investment in technology to radically reshape the company's supply chain infrastructure to implement CPFR and to provide high-quality products and incomparable service to customers. The $20 million investment in Manugistics, a CPFR software solution, enables Perdue to collaborate and share critical information on forecasting, point-of-sale data, promotion activities, inventory, and replenishment plans with its partners. Using Manugistics forecasting software and supply chain planning tools, Perdue has become more adept at delivering the right number of poultry products to the right customers at the right time.

The third week of November is usually Perdue's busiest time of year. However, with the assistance of Manugistics, the company's output does not change drastically. The big difference in November is the form that turkeys take. Most of the year, turkey is prepared for food parts and deli meats, while this time of year it is whole birds. Getting turkeys from farm to table is a race against time, so Perdue has turned to the CPFR technology solution to make sure its products arrive fresh. Each of its delivery trucks is equipped with a global positioning system that allows dispatchers to keep tabs on the turkeys en route from each of the company's four distribution centers to their destinations. In case of mechanical failure (e.g. flat tires), a replacement will be sent to rescue the palettes of poultry. "We know where our trucks are exactly at all times," says Dan DiGrazio, Perdue's director of logistics.

Perdue collaborates with major food service and grocery companies such as Chick-Fil-A restaurants, Wal-Mart, and Sam's Club for collaborative production planning, demand forecasting, and inventory replenishment. In implementing CPFR, Perdue assesses its production capacity and makes necessary investments to satisfy the expected demand level. In 2004, Perdue purchased a 500,000-square-foot plant near Perry, Georgia, from Cagle's for $45 million. The plant processes up to 350,000 birds per week to meet the demand from Chick-Fil-A restaurants, and an additional 450,000 birds per week for tray-pack products sold at Wal-Mart and Sam's Club.

Vision 2020

Perdue Farms is a rapidly growing company with a bold vision: "To be the leading quality food company with $20 billion in sales in 2020." Perdue Farm's vision of year 2020 is to increase sales volume to $20 billion from its current level of $2.7 billion.

The market needs for chicken are growing, as low-carb diets are becoming more popular. Perdue's products are the answer for health-conscious customers, with less fat and fewer calories than other meat products. Perdue products are a good source of protein as well.

Tyson is a strong competitor of Perdue. In general, a grocery store carries three brands of uncooked chicken products. Many national grocery chains, such as Kroger, Farm Fresh, and K-Mart, carry both Tyson and Perdue brand products. Other poultry producers, such as Sanderson Farms and Gold Kist, are competing for third place.

Today, Perdue is further expanding its production facilities and product variety. Facing the market opportunities and competition, the management of Perdue is thinking about strategies and methods that will help the company to realize its Vision 2020.

Discussion Questions

1. Have you consumed Perdue Farm's products? Which one of Perdue's products do you like best? Which one do you not really care for?
2. Describe the production process at the chicken processing facility.
3. Is the vertically integrated supply chain that Perdue Farms operates the best option for Perdue? Why or why not?
4. Compare the poultry industry's vertical supply chain model with Dell's direct model that outsources most of its operations. Is this the best operations management/supply chain model for both industries? Why or why not?
5. Analyze the entire production lead-time, starting from eggs. How does lead-time affect production planning? Are there any risks and bottlenecks?
6. Discuss Perdue's strengths, weaknesses, opportunities and threats. What new operational ideas and changes would you recommend to Perdue management to realize its vision 2020?

References

enVISION (2005). Manugistics Forum Keynote Speakers, Atlanta, GA. USA, retrieved August 30, 2005 http://www.manugistics.com/envision2005/speakers.html.

George, M., (2005). "Perdue to add cooking plant," Houston Home, June 17, 2005, retrieved August 30, 2005, www.news.mywebpal.com/partners/963/public/news640284.html.

Lanter, Charlie (2004). Perdue Farms Executives Show Off Perry, Ga., Facility, American Stock Exchange, May 13, 2004, retrieved August 30, 2005, www.amex.com.

Luttrell, Sharron Kahn, Talking Turkey with Perdue's CIO, CIO Magazine, Nov. 1, 2003, retrieved August 30, 2005, http://www.cio.com/archive/110103/tl_scm.html?printversion=yes.

Websites

www.ext.vt.edu/cgi-bin/WebObjects/Docs.woa/wa/getcat?cat=ir-lpd-pou

www.osha.gov/SLTC/etools/poultry/followup.html

www.perdue.com

www.thepoultrysite.com/

2

Supply Chain Risk Management

Case 3

Improving Stanford Blood Center's Platelet Supply Chain[1]

Yenho Thomas Chung[†], Feryal Erhun[‡], and Tim Kraft[°]

William leaned back in his chair and looked at the platelet outdate numbers for April again. One-third of the platelets outdated—how could this be? William's boss and the platelet donors would not be pleased. As the project manager for Stanford Blood Center's (SBC) process improvement initiative, William knew that he and his team had to do something fast; otherwise, SBC would start to lose valuable donors and incur platelet shortages. Due to the short shelf-life for platelets (five days) and volatile demands, it was critical that SBC carefully manage the platelet supply process, which included distributing platelets to local hospitals as well as collecting and redistributing unused platelet inventory among those hospitals. However, as April's numbers demonstrated, there were significant flaws in SBC's platelet supply process.

Because his team was mainly composed of medical personnel and staff, William decided to seek outside support for SBC's platelet

[1] This case was prepared by Yenho Thomas Chung (LG CNS Entrue Consulting Partners), Professor Feryal Erhun (Stanford University), and Professor Tim Kraft (University of Virginia) based on field research of an actual business situation. Some names, dates, and data are disguised, and some material is fictionalized for pedagogical reasons. It was written as a basis for class discussion rather than to illustrate effective or ineffective handling of an administrative situation. The authors would like to thank their collaborators at Stanford Blood Center and Stanford University Medical Center for their efforts.

[†] LG CNS Entrue Consulting Partners, Seoul, South Korea; yhchung@lgcns.com

[‡] Stanford University, Stanford, California, USA; ferhun@stanford.edu

[°] University of Virginia, Charlottesville, Virginia, USA; KraftT@darden.virginia.edu

process problem. A few months earlier, he had read an article in the Stanford Report on risk management in supply chains by Professor Feryal Erhun from Stanford University's Department of Management Science & Engineering (MS&E). William recognized many of the symptoms that Prof. Erhun discussed in her article in the SBC supply process. Now he just hoped this afternoon's meeting with Prof. Erhun and one of her Ph.D. students would provide insights into SBC's supply problem.

Stanford Blood Center: "Give Blood for Life"[2]

Located in Palo Alto, CA, Stanford Blood Center (SBC) is a not-for-profit organization, which was established in 1978 to meet the increasing needs of Stanford University Medical Center (SUMC), which comprised Stanford University Hospital and Lucile Salter Packard Children's Hospital. Since then SBC has expanded its scope to serve El Camino Hospital, the Palo Alto VA, and O'Connor Hospital. In addition to providing blood testing and transfusion services, SBC also acts as a teaching and research setting for Stanford medical students and faculty. It is the second-largest transfusion facility in the U.S with approximately 60,000 blood donations and 100,000 blood products for medical use per year.[2]

SBC has always been on the forefront of transfusion medicine. For example, in 1983, two years before the AIDS virus antibody test was developed, SBC became the first blood center to screen for AIDS-contaminated blood. In 1987, SBC became the first blood center in the U.S. to screen donors for HTLV-I, a virus believed to cause a form of adult leukemia. Additionally, SBC was the first blood center in the world to routinely test for cytomegalovirus (CMV) and provide CMV-negative blood for immuno-compromised transfusion recipients, and SBC was among the first in the U.S. to provide human leukocyte antigen (HLA) compatible platelets.[2]

Although most medical institutions are known to be conservative and reluctant to share their information with outsiders, SBC has always

[2] http://bloodcenter.stanford.edu/

fostered a collaborative environment. This is why William sought out the help of the MS&E team. However, even with the support of the MS&E team, William knew that the sweeping improvements SBC desired could not be achieved without the cooperation of SBC's largest customer, SUMC. William had unsuccessfully approached SUMC staff previously about working together to improve the supply process. Without solid evidence that there were inefficiencies in the supply chain between SBC and SUMC, it was difficult to convince SUMC staff that collaboration between SBC and SUMC would improve the system efficiency. To get the SUMC staffs' attention this time, William would need tangible evidence of the inefficiencies in the current supply chain.

Platelets: Perishable Inventory with Five-Day Shelf Life

Platelets are small cell fragments found in the blood plasma of mammals. Platelets are responsible for starting the formation of blood clots when bleeding occurs and thus are often transfused to patients to treat or prevent bleeding during surgeries. There are no artificial substitutes for platelets; platelets transfused to a patient must be collected from another human being. In addition, the donor pool for platelets is limited due to restrictions on the number of times a donor can donate per year and the often lengthy donation process.

Platelets can either be isolated from whole blood donations or collected by an apheresis process that requires sophisticated instrumentation and a highly trained support staff. Whole blood donations are usually collected from random walk-in donors in mobile stations or local blood collection centers. Whole-blood-derived platelets are then split from a unit of whole blood that has not been cooled yet. A whole unit of blood contains not only platelets but also red blood cells, white blood cells, and plasma. Therefore, the amount of whole-blood-derived platelets from a single bag is typically not significant. In contrast, an apheresis device draws blood from a donor and centrifuges the collected blood to separate out platelets and other components. Because the remaining blood is returned to the donor during the process, an apheresis donor can provide more platelets than a whole-blood donor can. Typically, an apheresis donor can donate at least one therapeutic dose, whereas generating a therapeutic dose from whole-blood-derived platelets requires multiple donations and multiple donors. Because apheresis platelets come from a single donor, they are usually less risky in terms of transfusion-transmitted disease, especially since apheresis platelet donors are registered and closely monitored by the blood center staff. In addition, because an apheresis dose does not contain as many red blood cells as a whole-blood-derived dose, apheresis platelets do not have to be cross-matched in terms of blood types (i.e., ABO and Rh+/-).

One of the most challenging aspects regarding platelet inventory management is the extremely short shelf-life for platelets. Whereas some European countries allow 7 days of shelf life, in the U.S., Food and Drug Administration (FDA) regulation requires that every unused platelet unit be discarded after 5 days. In addition, due to the 48-hour testing process that is required right after the donation, the practical shelf life of platelets is actually only 3 days in the U.S. In part due to this short shelf life, 10.9% of apheresis platelet units collected in the U.S. went outdated in 2006; i.e., they expired without being transfused. Therefore, to maintain a reasonably high service level and a low outdate rate, the balance between the supply and the demand for platelets must be closely monitored.

SBC's Platelet Supply Chain

Similar to a typical blood supply chain, SBC's platelet supply chain starts with the collection of platelets from donors. Once collected, the platelets are then processed, tested, and delivered to the hospitals where they are either transfused to patients or outdated as shown in Figure 3-1. There are three critical processes within this supply chain: the collection process, the rotation process, and the issuing process.

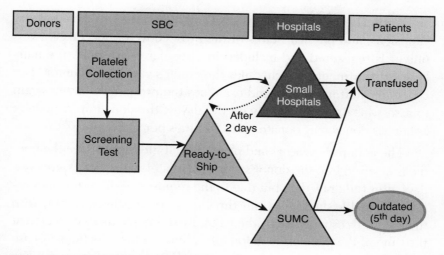

Figure 3-1 SBC's platelet supply chain.

The Collection Process: Platelet Collection and Testing

On March 1, 2004, the AABB[3] mandated that each blood center and transfusion service should implement methods to limit and detect bacterial contamination in all platelet components. To comply with this requirement, SBC transitioned to bacterial culture of platelet products and the exclusive use of apheresis platelets. Although SBC is well-equipped and well-staffed, collecting platelets through an apheresis process is a difficult task. While whole-blood-derived collection accommodates random walk-in donors, an apheresis process limits the effective donor pool to altruistic, consistent, and highly committed donors due to instrument immobility and the lengthy 1½–2½ hour collection time.

SBC's marketing department recruits volunteers for regular apheresis platelet donation. It is not easy to recruit a new apheresis platelet donor. The volunteers donate platelets without any compensation; thus, it is very important to maintain a low outdate rate so as not to impair donors' motivations. For new volunteers, the donor

---

[3] AABB is "an international, not-for-profit association representing [nearly 8,000 individuals and 2,000 institutions] involved in the field of transfusion medicine and cellular therapies" (http://www.aabb.org).

collection staff performs interviews and checks the medical history, vital signs, and physical status of the prospective donor. Once a volunteer is registered and included in a regular donor pool, the marketing department pre-schedules the donor's visit to the center. The collection staff interviews and examines the prospective donor again to assess his or her eligibility on the day of the donation. A platelet apheresis donor may donate up to 24 times per year.

The apheresis process and platelet shelf life begins once the needle is inserted into the donor's vein. The platelet product expires five days after the draw date but due to the requisite 48-hour testing process, the usable shelf life is only three days. All tests must comply with the standards established by the FDA. First, sample tubes collected at the time of the platelet donation are submitted for infectious disease testing (e.g., blood-borne agents such as HIV, hepatitis, and syphilis). Second, a sample of the platelet unit is cultured for bacterial growth (i.e., bacterial detection). The infectious disease testing is usually completed within 24 hours but the bacterial detection test requires 48 hours. Finally, platelets suitable for transfusion are shipped to the hospitals with up to 3 days shelf life remaining. Figure 3-2 demonstrates the platelet availability schedule based on draw date.

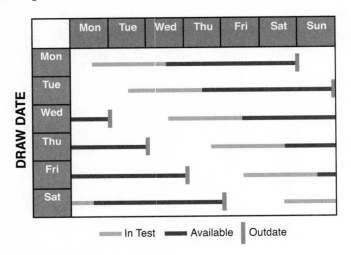

Figure 3-2 Platelet availability schedule.

Currently, SBC schedules donor visits based on the facility capacity, regardless of the incoming demand or the current inventory level. Thus, the number of units collected each day is fairly constant. In order to collect the pre-determined daily amount, registered donors are scheduled several days in advance (Monday through Saturday). Based on historical collection data, the average collection level per day is 40 units Monday through Friday and 60 units on Saturday. Because there is no collection on Sunday and donors are more readily available on Saturdays, SBC collects more platelet units on Saturday.

There are several uncertainties in the platelet collection process. For instance, donors may not show up due to personal emergencies, or some donors may not donate due to their own medical status on the day of the scheduled donation. Thus, to ensure that its needs are met, SBC schedules more donors than needed to protect against unexpected loss in supply as well as unexpected spikes in demand. One additional issue with the collection schedule is that some donors can donate up to three units during a visit due to individual attributes such as higher platelet counts and body surface. In order to cope with supply uncertainties, SBC often collects the maximum number of units from each donor. Consequently, on a given day, SBC may collect a larger amount of platelets than needed; even though this then increases the risk that collected units will go outdated. Despite these precautions, unexpected platelet shortages still occur. When this happens, SBC's own collection process typically cannot respond quickly enough to demand due to the two-day testing period. In such situations, SBC procures additional platelet units from other regional blood centers.

The Rotation Process: Supply Contracts and Platelet Rotations

SBC's largest customer, SUMC, is primarily supplied by SBC. SUMC transfuses approximately 9,100 units of platelets per year, which consumes about 80% of SBC's platelet supply. To utilize its remaining apheresis platelet capacity, SBC also serves small local hospitals. SBC works hard to market to these local hospitals. Although most blood centers are not-for-profit organizations, the blood product market is very competitive and generous contracts are often used to attract new customers. Due to this competition, SBC offers consignment contracts to the small local hospitals. Under a consignment contract, SBC delivers fresh platelets based on the hospital's daily demand. After two days, the hospitals can then send any leftover or unused units back to SBC. These units, with one day of shelf life remaining, are called "short-dated units." The small hospitals are not financially responsible for the short-dated units. They only pay for the units that they transfuse and any units that expire while still in their inventory. Consequently, in order to avoid any financial responsibility, the small hospitals return almost all unused, two-day-old units. For the small hospitals, this contract is attractive because they incur very little risk. Conversely, SBC incurs high risk since it does not know how many short-dated units a small hospital might return. When the short-dated units are returned to SBC, these units are rotated to SUMC along with fresh units. Short-dated units are sent to SUMC, because it has the highest demand among the hospitals and, therefore, provides the best opportunity for the short-dated units to be transfused before the end of the day.

SUMC orders platelets from SBC two to four times daily. The ages of platelet units delivered to SUMC differ, ranging from three or four days (fresh units) to five days (short-dated units) old. Although SBC deliveries are triggered by demands at SUMC, not every unit delivered is added to SUMC's available inventory. On rare occasions, platelets are discarded due to cancelled transfusions, being out of refrigeration too long, or to blood center requests to return questionable products (e.g., donors providing post-donation information that affects eligibility).[4] The major concern for SBC and SUMC, however, is outdated units. As there is little consistency in the age of the platelets SUMC receives, there is a high potential for many of the units to become outdated. Due to this high perishable inventory risk, SBC and SUMC have a cost-sharing contract on the outdated units, which is shown in Table 3-1.

Table 3-1 SBC and SUMC Cost-Sharing Agreement

Age of Platelets When Shipped	Cost Sharing if Outdated
4 days old or younger	SBC (50%)–SUMC (50%)
5 days old (short-dated unit)	SBC (100%)–SUMC (0%)

---

[4] Since the number of discarded units is negligible, they are not included in the data set provided.

According to an AABB survey report, U.S. hospitals paid an average of $538.72 for a unit of apheresis platelets. This cost includes procurement costs, operations costs, testing costs, and processing costs. As shown in Table 3-1, SUMC incurs 50% of the costs if it receives the product before the day of expiration (i.e., fresher units) and does not transfuse it. Alternatively, SUMC incurs no costs for outdated units when SBC ships short-dated units. SBC staff believe that this cost-sharing contract reduces the overall system outdate rate, since SUMC receives short-dated units as well as fresh units because of the pricing structure of the cost-sharing contract.

The Issuing Process: Platelet Demand in SUMC

SUMC utilizes platelets for many different types of operations, ranging from day-to-day operations to pre-scheduled or emergency surgeries; therefore, SUMC always has to stock a certain amount of platelet units in its inventory. The platelet supply chain between SBC and SUMC is completely decentralized; i.e., platelet collection and rotation decisions are made solely by SBC and platelet ordering and issuing decisions are made solely by SUMC. From SUMC's previous ordering and issuing history, William identified that SUMC orders on average 32 platelets per day but transfuses only 25 platelet units on average per day.

A Snapshot of the SBC-SUMC Supply Chain in 2006

For their analysis, William and his team collected data for the number of transfusions and outdates during the last four months, from January 2006 to April 2006. Because obtaining this information from the small local hospitals SBC served would be difficult, the team focused solely on SUMC transactions. The data set includes unit number, draw date by SBC, expiration date, received date by SUMC, issue date by SUMC, and if the unit is outdated instead of transfused, the outdate date. Figure 3-3 demonstrates the outdate rate at SBC and the number of transfusions at SUMC from January 2006 to April 2006. Notice that in April, the outdate rate increased to 30+%, compared to 15% to 20% during the previous months. At the same time, there was a significant decrease in demand in April.

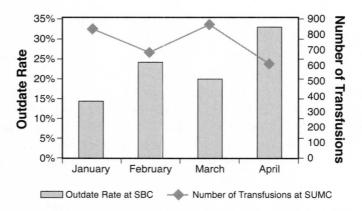

□ Outdate Rate at SBC   ◆ Number of Transfusions at SUMC

Figure 3-3 SUMC and SBC outdate rate and number of transfusions (Jan.–Apr. 2006).

The Problem

After hearing William's overview of the platelet supply chain, the MS&E team understood the significance of the problem. Although SBC was concerned about increasing costs, the more important issue was the donors. Maintaining a large donor base was extremely important to SBC. Because every donor wants his or her donation to be transfused rather than perish, if SBC continued to maintain a high outdate rate, it risked losing valuable donors. William and the MS&E team agreed to meet in two weeks, after the MS&E team had time to review the data and their notes.

As he led the MS&E team to the lobby of SBC, William thought again about the platelet supply process. What could be causing the high outdate numbers? Could it be the agreements SBC had set up with the hospitals? Was it the collection process itself? Or were there hidden inefficiencies in SUMC's transfusion process? William hoped that a solution could be found.

Questions

After listening to William's description of the platelet supply chain, the MS&E team has come up with the following list of questions to answer over the next two weeks:

1. Based on the data provided, are there any potential imbalances between the existing demand and supply patterns?

2. Identify the characteristics of outdated items at SBC/SUMC. Is there any connection between these characteristics and any of the three processes discussed (i.e., collection, rotation, and issuing)?

3. To decrease the outdate rate of platelet units, what recommendations should be made to improve the platelet supply chain between SBC and SUMC? Is there any incentive for either party to follow these recommendations?

4. Identify at least one problem in each processing area (collection, rotation, and issuing), and provide a solution for each of them.

Case 4

Financial and Operational Risk Management at Molson Coors[1]

Dennis Kira[†], Ahmet Satir[‡], and Dia Bandaly[°]

Company Background

Formed by the merger of Molson Inc. and the Adolph Coors Company in 2005, the Molson Coors Brewing Company (the "Company") is the fifth largest brewer in the world by production volume. The Company brews and sells about 40 different beer products, in addition to selling beer via partnerships with companies like Heineken and Corona. Molson Coors thrives particularly in Canada, where it commands over 42% of the market, largely through sales of its flagship brands Coors Light and Molson Canadian.

In 2007, the Company announced a joint venture with SABMiller (LSE: SAB). The deal, forming the second-largest brewer in the U.S.

[1] The authors thank the executives and managers of Molson Coors for their full cooperation during the write-up by providing enterprise specific information and data, as well as verifying the final text in a thorough manner to ensure factual validity and data confidentiality. The authors express their gratitude to Export Development Canada (EDC) for initiating the idea for the case subject, providing feedback as to the context and content and sponsoring the case. All figures provided are distorted in a proportional manner to protect the confidentiality of the Company's data. Unless otherwise stated, all monetary figures are in U.S. dollars.

[†] Concordia University, Montreal, Quebec, Canada; dkira@jmsb.concordia.ca

[‡] Concordia University, Montreal, Quebec, Canada; asatir@jmsb.concordia.ca

[°] Concordia University, Montreal, Quebec, Canada; dbandaly@jmsb.concordia.ca

after Anheuser-Busch Companies (NYSE: BUD), was completed on July 1, 2008. The joint venture encompasses only the Company's operations in the U.S.; its Canadian and UK businesses remain completely under the control of Molson Coors. The combined Company benefits greatly from logistical and transportation synergies realized through the joint-venture. The Company expects to realize ultimate cost savings of $500 million.

In 2009, Molson Coors posted a net income of $720.4 million. The increase in net income is attributable partly to $92 million of cost savings as part of its now-completed Resources for Growth (RFG) cost savings program. Over the past three years, the Company has delivered $270 million in cost savings through the RFG program, which significantly exceeded the Company's commitment of $250 million. MillerCoors also delivered incremental cost savings of $26 million in 2009, which are part of its $200 million of second-generation cost savings that are expected to be delivered by the end of 2012.

In a press release on the financial performance of the Company in 2009, Peter Swinburn, President and CEO of Molson Coors, stated: "Overall consumer demand remains sluggish, and we see these conditions continuing to impact volume and mix in the near term. Our strategy remains consistent, however. We are focused on investing in innovation and in our brands and ensuring we maintain a strong balance sheet, so that when market conditions improve we are better positioned to accelerate our growth and capitalize on opportunities. Looking to 2010, we expect volume to remain challenging, especially in the first half, but we are focused on continuing to establish a strong brand base to our business that ensures we not only manage the current market but that we take full advantage of revenue upsides when momentum improves."

Of the 40 different brands of beer sold worldwide (14 brands in the U.S.), the bulk of the Company's sales volume is concentrated in three different products across three countries: Coors Light in the U.S., Carling in the U.K. and Coors Light and Molson Canadian in Canada. Therefore, consumer preferences for these three products alone have a significant impact on the success of the Company. Especially in the U.S., Molson Coors competes with small, local breweries for consumers' tastes and preferences. There are approximately 1,500 small breweries in the U.S., and on average, each produces 5,600 barrels of beer a year. Although this pales in comparison to the 70 million barrels that MillerCoors produces in the U.S. alone, the Company must overcome staunch loyalties when marketing products in regions with strong local breweries.

Similar to all brewers, Molson Coors is exposed to raw materials costs as part of the brewing and packaging process. Not unlike many commodities, prices for the most important input materials, aluminum, barley, and grain, fluctuate widely. For example, aluminum prices have fallen more than 60% from their 2008 highs of $3,300/metric tonne to less than $1,430/metric tonne, although in 2010 the price of aluminum had recovered to above the $2,000/metric tonne threshold.

Competition

In 2009, Molson Coors was the fifth largest brewer in the world, as shown in Exhibit 4-1. Its sales are focused in three countries:

United States: Beer is the most preferred alcohol in the United States, with 42% of alcohol drinkers choosing it. Wine is the second most preferred, with 31%, followed by hard liquor at 21%. Although beer has been the historical preference of Americans, Molson Coors still has to compete with these other categories of alcohol. The MillerCoors joint venture is the second largest brewer in the U.S., with 29% of the market. The company trails its larger competitor Anheuser Busch owned by InBev (INB-BT), which has 49% of the U.S. market.

Canada: Molson Coors is the largest brewer in Canada with 42% of the market share by volume, although Labatt Breweries of Canada (owned by the world's largest brewer, ABI) is only a few percentage points behind. Of the two major brands, Coors Light has about 15% of the market share and Molson Canadian has about 10%. Canada is a mature market that is characterized by heavy competition among large-scale producers, regional breweries and microbreweries.

United Kingdom: Coors Breweries Limited is Molson Coors's arm in Western Europe. It has an approximate 25% market share of the British market, Europe's second largest market.

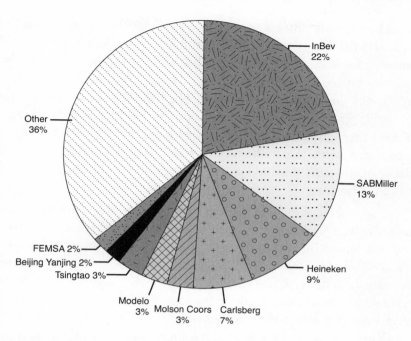

Exhibit 4-1 World beer market share by volume in 2009.

The Brewing Process

Brewing the perfect beer requires the brewer to use art, craft and science, in a balance of natural ingredients and processes. Some brewers embrace modern technology, while others use more traditional means. However, whether the brewery is large or small, old or new, the brewing process remains the same. The brewery industry includes over 10,000 breweries with combined annual revenue of over $50 billion worldwide. The major driver of demand is consumer leisure activity. The profitability of individual companies depends on marketing, distribution and operational efficiency.

Major brewery products are malt beverages, primarily beer and ale, packaged in cans, bottles, barrels or kegs. In Canada, canned beer accounts for about 35% of industry revenue; bottled beer for about 55% and kegs for about 10%. Additional products include other malt beverages, such as porter, stout and non-alcoholic beer, and brewing materials, such as brewers' grains and malt extracts.

The brewing process takes two to three weeks depending on the product. Breweries crack purchased malts by milling and then add water to form a mash, a mixture of hot water and crushed grain. The mash is heated and stirred in a mash tun (a large cask for liquids) to convert the mixture into fermentable sugars. The mixture is then strained and rinsed in a lauter tun to produce wort, a liquid with high levels of fermentable sugars. The wort flows from the wort receiver into a brew kettle that boils and concentrates the liquid. The resulting flavor of the wort depends on the hops additives, temperature, and length of brewing.

The next steps include straining, cooling, and storing in a fermentation cellar. Brewers add yeast to jump-start fermentation, which converts sugars into alcohol and carbon dioxide, the source of carbonation. The fermented beer cools for about a week until it clarifies and develops the desired flavor. Filtration, if used, removes extra yeast, after which the brew is ready to package for delivery to distributors. Breweries package beer in bottles or cans, typically in 6 or 12 packs, for delivery in cases for eventual retail sale and in barrels or kegs for on-premise draft sales. The brewing process is illustrated in Exhibit 4-2. The key beer industry production metric used globally for volume is measured in the number of hectoliters (HLs) a brewery produces or sells per year.

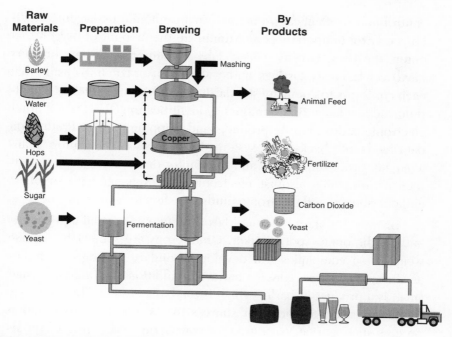

Exhibit 4-2 Brewing process.

It is typical for large brewing companies to strategically locate their breweries near major population centers to minimize the shipping costs associated with their finished products. Molson Coors has five breweries and 40 distribution centers within Canada. Within the Canadian market, the retail sales model for beer varies considerably by province, as alcohol sales are provincially regulated in Canada. The models include beer being sold through provincial liquor stores (Nova Scotia, Saskatchewan), within the grocery store or convenience store channel (Quebec, Newfoundland), in private liquor stores (Alberta) or in stores dedicated to the sale of beer (Ontario). Based on the "go to market" model that exists, breweries may alter their distribution model accordingly as well. For example, in the province of Quebec, each major brewery has established its own distribution operation for the shipment of beer to the retail and on-premise customer, while in the Western Provinces, Labatt and Molson have formed a joint-venture to distribute beer called Brewers Distribution Limited (BDL).

The brewery business is highly automated. Advanced process equipment and filtration systems monitor each batch to flag quality control and mechanical problems. Environmental management systems control temperatures and minimize the amount of oxygen that enters the beer. Quality control labs are important. Some brewers have over 125 tests, tastings, and evaluations per batch to ensure that each conforms to company standards. Breweries use automated bottling and keg lines. Radio frequency identification (RFID) and other electronic codes identify products and shipping pallets. Production data feeds into back-office systems for analysis and inventory management, order fulfillment, and to monitor distributor sales commitments. Companies also use electronic data exchange with suppliers and distributors and electronic funds transfers to receive payments.

Breweries obtain raw materials through contractual agreements and on the open (spot) market. Grain crops are subject to adverse weather, so companies will develop secondary geographic sources as alternatives, especially for barley. In addition to purchasing malt (malt is a product of barley which is the basis of beer), brewers may purchase various other sugar sources to assist in the fermentation process such as rice, corn grits, or corn syrup. Packaging materials used by brewers include corrugated paper boxes, paperboard boxes, aluminum cans, bottles, labels, crowns, and kegs. The input costs for many of these materials, such as aluminum used for can production, can be volatile, and therefore brewers will look for ways to manage this volatility. Tools to manage this volatility in this case study include long-term supply contracts and commodity and currency hedging to help manage supply and costs. This study focuses on the financial and operational risk management techniques used for hedging in the context of aluminum cans.

Financial Risk Management Background

The corporate world has hedged its costs and revenues for decades. Through futures, forwards, options, and swaps, companies have hedged risks related to stock investments, commodities, interest rates, currency, and relevant indexes. A common feature for these types of risk is that the risks are mainly related to price. A "derivative" is defined as a financial instrument that has a value determined by the price of "something else." What is described as "something else" is more commonly called the "underlying asset." Before expiry, other factors like time to expiry, volatility of the underlying asset, and expected development contribute to determine the value of a derivative. A derivative has an expiration date where the derivative ceases to exist. At that point, the value of the derivative is entirely determined by the price of the underlying asset.

Effective hedging requires a clear understanding of the relation between the hedged position and the hedging instrument. The strength and direction of the linear relation between two variables may be measured with the use of covariance and correlation statistics.

In order to assess costs and benefits, it is crucial that the company has expertise that understands the derivatives it is trading. Such expertise may come at a high cost through, for example, highly educated employers or expensive consulting firms. Derivatives also have implications after they are traded. Transactions need to be monitored to evaluate how the hedge is performing. Furthermore, derivative transactions have tax and accounting consequences. In particular, derivative transactions may complicate financial reporting. This might be both time consuming and costly.

Payoff on a derivative depends on the price of the derivative's underlying asset. If an asset and the underlying asset of a derivative are perfectly correlated, there is no basis risk. Basis risk arises as soon as an asset and the underlying asset of a derivative are not perfectly correlated. This imperfect correlation between the asset and the underlying asset of the derivative creates potential for excess gains or losses in a hedging strategy. Imperfect correlation reduces efficiency of the hedging instrument and increases risk of the total portfolio.

Commodity futures have been widely used as a risk management tool. A commodity future reduces risk by locking a future price, thereby removing price risk. As a result, if an energy reseller experiences a "normal" winter, a commodity future will work properly. On the other hand, should the winter be abnormally warm, demand for energy will fall. As a result, the energy reseller's revenues will decline. The commodity future hedge will probably work partly as energy prices tend to fall during a warm winter. However, the commodity future does not protect against the low demand, and the energy reseller may experience low revenues even though commodity futures were used to hedge risk.

For this reason, risk managers hedge non-core risks like foreign exchange, interest rates, commodities, equities, credit, natural catastrophes and, nowadays, the weather. The focal goal of risk management is to increase shareholder value. Shareholders prefer a less volatile earnings stream to a volatile one. Therefore, companies that minimize earnings volatility, mainly through removing non-core risk, accomplish higher equity multiples, stronger credit ratings, lower cost of debt, and improved access to funding.

Financial Risk Management at Molson Coors

Molson Coors utilizes a variety of agricultural and commodity products in brewing and bottling/canning its beverages. For beer, the most important inputs are barley and hops. Barley typically constitutes 15% of the brewing costs of beer, and a significant price increase in barley, for instance, would increase the cost of the company's goods sold and put pressure on margins. During 2007–2008, barley prices almost doubled because of dwindling supply caused by consecutively poor harvests and increasing global demand. Further pressure on barley prices has arisen since farmers are increasingly attracted to farming crops such as corn and soybeans instead of barley because of the burgeoning bio-fuel industry. During the commodities super spike of 2007 and 2008, the prices of these commodities rose drastically with the general commodities bubble and dramatically pressured beer company margins. They receded in late 2008, but remain at historically elevated levels. The possibility of another significant rise in commodities represents a constant threat to profits for beer companies globally.

For commodities traded internationally, the strength of producers' and consumers' currencies can affect the prices of commodities. For example, even if Brazil (the world's leading sugar producer) produces an abundance of sugar in any given year, sugar prices will probably remain inflated if the Brazilian Real is particularly strong relative to other currencies. Then, currency exchange rate hedging, such as currency options, provides further protection. Molson Coors records its financial information in U.S. dollars for corporate reporting but realizes significant portions of its revenues in other currencies such as the Canadian Dollar and the British Pound Sterling. Like other multi-national companies, Molson Coors is naturally affected by the currency fluctuations for its revenue and cost reporting, as it trades its raw materials and products globally.

There are two committees at Molson Coors that deal with financial risk management: the Commodity Risk Management Team (CRMT) and the Financial Risk Management Committee (FRMC). The former, as its title suggests, largely deals with hedging against commodity (such as barley, corn, aluminum, energy) risks. Although currency exchange risk is considered by this committee as well, final decisions as to hedging against this risk are made by FRMC, which has a wider mandate in financial risk management. The composition of CRMT involves corporate team members (assistant treasurer, financial risk management manager, senior financial analyst) and members from local procurement (strategic sourcing manager, senior financial analyst). The FRMC is composed of a much broader group, including the chief financial officer, procurement officers, global treasurer, global controller, global assistant treasurer, financial risk management managers, and senior financial analysts.

A sample agenda for a typical CRMT meeting is given in Exhibit 4-3. Excerpts from a CRMT meeting are also provided below to give a sense of discussions that ensue in such meetings. Forward contracts data are provided in Exhibit 4-4 for the October 2008–December 2010 period. Exhibit 4-4 entries are aluminum futures long hedge positions where, for example, trade date 08/31/08 with exposure date Oct. 08 indicates that on 08/31/08, the Company took a long position for 200 metric tonnes of aluminum forward contracts to be exercised in October 2008. Exchange rate hedging data are presented in Exhibit 4-5 for the year 2009. An entry such as 01/15/09 with FX rate $1.0134 in Exhibit 4-5 indicates that the currency hedge position entered by the Company, at different trade dates within the previous two years, has a maturity date of 01/15/09 for $7.5 million to meet the payment of $9,208,859 at the average exchange rate of 1.0134. Exhibit 4-4 is the result of deliberations by the CRMT, whereas the data in Exhibit 4-5 originate from FRMC decisions.

CRMT Meeting–Agenda

• Market outlook summary

• Review current hedged positions–Summary

    • Annual Spend, Current Coverage, Price, Budget

• Market summary for each commodity position

    • Aluminum, Natural Gas, Barley, Corn

• Supplier Financial Risk Review

Exhibit 4-3 Typical agenda of a CRMT meeting

Exhibit 4-4 Forward aluminum contracts for October 2008–December 2010

Exposure Month	Trade Date				
	08/31/08	09/14/08	10/27/08	01/06/09	03/26/09
Oct-08	200				
Nov-08	190				
Dec-08	250				
Jan-09	100	75	75		
Feb-09	80	65	65		
Mar-09	150	90	90	90	
Apr-09	130	140	140	140	
May-09	110	165	165	165	165
Jun-09	150	250	250	250	250
Jul-09	50	350	350	350	350
Aug-09	90	110	110	110	110
Sep-09	80	90	90	90	90
Oct-09	40	65	65	65	65
Nov-09	70	65	65	65	65
Dec-09	60	65	65	65	65
Jan-10	40	40	40	100	100
Feb-10	40	40	40	100	100
Mar-10	90	40	40	100	100
Apr-10	90	40	40	100	100
May-10	90	40	40	100	100
Jun-10	90	40	40	250	250
Jul-10	55	40	40	250	250
Aug-10	50			250	250

Exposure Month	Trade Date				
	08/31/08	09/14/08	10/27/08	01/06/09	03/26/09
Sep-10	40			150	75
Oct-10	20			75	50
Nov-10	20			75	50
Dec-10	20			75	50
Total (Metric Tonnes)	2,395	1,810	1,810	3,015	2,635
Midwest Transaction Price (USD)	$2,660	$2,400	$1,952	$1,744	$1,605

Exhibit 4-5 Exchange rate hedging data for the year 2009.

Maturity Date	Average of FX Rate	Sum of Trade Amount (USD)	2009 Actual Exposure (USD)	Hedged Percentage
01/15/09	$1.0134	$7,500,000	$9,208,859	81.5%
02/17/09	$1.0098	$7,600,000	$8,224,438	92.4%
03/16/09	$1.0132	$9,100,000	$12,028,004	75.7%
04/15/09	$1.0089	$9,500,000	$12,312,554	77.2%
05/15/09	$1.0092	$12,400,000	$14,796,188	83.8%
06/15/09	$1.0352	$14,400,000	$19,792,246	72.8%
07/15/09	$1.0066	$12,900,000	$13,334,098	96.7%
08/17/09	$1.0101	$11,100,000	$13,647,706	81.3%
09/15/09	$1.0556	$9,800,000	$11,597,687	84.5%
10/15/09	$1.0351	$7,200,000	$9,098,632	79.1%
11/16/09	$1.0593	$9,150,000	$10,509,517	87.1%
12/15/09	$1.0198	$7,600,000	$10,372,095	73.3%
Averages and Totals	$1.0240	$119,250,000	$144,922,024	82.3%

Excerpts from a CRMT Meeting

Discussions focused on how to manage volatility of company earnings and cash flows due to exposure on commodity price fluctuation in the future. More specifically, the members dealt with possible use of commodity swaps and commodity forward contracts. Techniques such as VaR (Value at Risk) and its use in the assessment of hedging decisions for each individual commodity attracted much heated discussion.

Following these discussions, the strategic sourcing manager stated:

> In recent months, we've all grown accustomed to a mixed bag of commodity prices. While there are important trends, risks and opportunities to be gleaned from economic changes, I don't want to be faced with drastic changes without some protection. So, the main question we need to address is: "We have few strategies that we can follow based on the past experience on commodity hedging, but we are not sure as to the best strategy to follow at this time." So prior to finalizing our strategy on barley and aluminum hedging for next year, I need some inputs concerning this matter.

At this point, the assistant treasurer indicated:

> The global boom in commodity prices in 2008—for everything from coal to barley—was fueled by heated demand from the likes of China and India, plus unbridled speculation in forward markets. That bubble popped in the closing months of 2008 across the board. As a result, farmers are likely to face a sharp drop in crop prices, after years of record revenue. Other commodities, such as aluminum, are also expected to tumble due to lower demand. This will be a rare positive for manufacturing industries, which will experience a drop in some input costs, partly offsetting the decline in downstream demand. Aluminum futures have settled below the symbolic $2,000/metric tonne mark on the London Metal Exchange for a full week of trading, suggesting that weak fundamentals might finally be catching up with the speculation-driven market.

To summarize, the senior financial analyst stated:

> Though price levels have in general dropped from their peak, volatility remains high and, as you are all aware, we have experienced hedging outcomes in the past that are as unexpected as they have been painful. Thus, we are under greater scrutiny from the Board, and complacency is not an option.

We need to be able to achieve reliable, predictable raw material costs. Hence, for commodity procurement and hedging, we need to improve our efforts in bringing financial and treasury expertise into the procurement through creating a cross-functional team so that we can avoid the danger of physical purchasing and financial hedging decisions made in separate silos. Such an approach would minimize the potential for creating unexpected inventory and financial positions, as well as increasing basis risk.

The chair of the committee then indicated:

Why don't we proceed to establish an approved hedge profile to be maintained by traders and as a start we can develop a commodity spend baseline and quantify the risk exposure. We can follow up with performing scenario and sensitivity analysis and explore macro medium- and long-term trends that impact market fundamentals. We should also analyze the terms of contracts with our commodity suppliers to determine whether any revisions need to be incorporated to reflect our revised hedging practices. We can consider using VaR to quantify the trade-offs between long-term contracts, spot prices and financial hedging to manage supply and demand levers and assess financial capabilities. If the members have suggestions or comments prior to establishing our commodity hedge positions, then kindly submit your comments within the next two days to me.

Operational Risk Management

Molson Coors operates five breweries in Canada. The three larger breweries are in Ontario, Quebec, and British Columbia, and the two smaller breweries are in Eastern Canada. The aluminum cans that are the subject of risk management in this case are supplied by two suppliers, one in Ontario and the other in the western U.S. The same suppliers also supply aluminum cans to other beer and soft drink companies. Since space is limited at breweries, aluminum cans are shipped by the supplier to off-site warehouses located near the breweries or directly to the brewery on a just-in-time basis. Aluminum cans constitute about 24% of unit material cost of producing a can of beer.

Forecasting

Full Goods (FG) forecasting is conducted using a demand planning forecast package used by the Company that contains up to four years of weekly sales history by SKU/warehouse and produces 36 months (156 weeks) of forecast. The application provides several forecasting methods, among them: Box-Jenkins, exponential smoothing, simple moving average, and regression. It allows demand planners to manipulate the forecast by changing parameter settings, changing (correcting) history and direct volume adjustments as required in order to get the best fit. Most forecasting is done at the lowest level (SKU/warehouse), although the application allows adjustments at other levels, such as SKU/key accounts and SKU/province, which are useful for adding promotional activity volume. The most common method used is exponential smoothing. "Specified smoothing option" available is utilized that allows the user to specify the parameters (such as the smoothing constant) used in forecast calculations. This method is used extensively, since it has provided reliable forecasts in the past with "acceptable" forecast errors.

Independent demand forecasting is a weekly process beginning on Monday when the application has been loaded with the previous week's sales or shipments and a new SKU/warehouse forecast has been calculated. Whether to enter sales or shipments differs by province. Demand planners review the forecasts and make adjustments where necessary, such as incorporating any new promotional activity. The final forecast figures, which are a combination of model outcome, intuition, and experience, are uploaded to SAP on Friday to be used for production planning, scheduling, and material requirements planning (MRP).

The weekly forecasts are also uploaded to a data warehouse program for reporting and analysis. The rationale for updating forecasts weekly is that the beer business is highly competitive and the product has a relatively short shelf life (around 180 days). Less frequent updating, say on a monthly basis, would substantially increase the risk of stockouts or ending up with obsolete products. By tweaking the forecast weekly, a more stable demand line is sent to allow for timely adjustments to production plans.

Material Requirements Planning (MRP)

MRP is run nightly in SAP. A weekly update is provided to the supplier with a 52-week projected demand by material. Materials are ordered on a weekly basis. The delivery schedules are requested by date and time for materials.

Procurement

The evaluation of a supplier to initiate supply is based on a series of criteria around quality, service, and cost. For cans, an audit of potential supplier facilities would be part of the supplier qualification process. Quality Concern Reports are kept for ongoing assessment of supplier performance. Innovation would also be a part of the service evaluation of suppliers. One can cite thermo activated Coors Light cans in this context. The ink on parts of the can indicates when the beer is cold enough to drink by changing the color of the can when the pre-specified degree of coldness is achieved. One other criterion in evaluating suppliers is cost. The cost performance of suppliers would be evaluated based on a cost model developed over time in house for that specific category.

Planning for can procurement lot sizing is done daily at the pallet level. Lot sizes are typically based on full pallet quantities by material type. The quantity to be purchased is driven by the daily MRP output and then finalized based on the production schedule for that day. Procurement quantity is then rounded up to the pallet quantity.

Inventory levels at Molson Coors are dependent on material type. Cans are typically brought in just-in-time (daily for that day's production due to space constraints at breweries). Off-site inventory levels are specified in purchase contracts for each material type and limited to a maximum of three months. However, inventory levels for cans at the off-site warehouses are typically at two to four weeks. The cans in these warehouses are owned by the supplier.

Transportation

All cans are shipped into Molson Coors breweries via trucks except for shipments made to the St. John's brewery, which are shipped via containers from the Port of Montreal to St. John's. The latter accounts for a very small portion of the overall can volume at Molson Coors.

For deliveries into the Vancouver Brewery, shipments are made using a 3PL provider who contracts on behalf of the Company with a number of different carriers. These shipments are brought from the suppliers' facilities to a staging warehouse off-site. Full truckloads are then shunted into the brewery as required.

Shipments to Toronto are primarily made using a back haul from the Company's own fleet for deliveries into the Toronto Brewery. The trucks deliver full goods outbound and then pick up can orders on their way back to the brewery. An off-site warehouse is also used to store some of the 473 ml cans and ends with longer lead times. However, the vast majority of the volume is processed using the back haul scenario from the suppliers' location.

Exhibit 4-6 provides the usage forecast and shipment data for small (355 ml) and large size (473 ml) cans for a (representative) peak period (July–August 2009) and a slow period (January–February 2010). Usage forecast figures are transmitted from Molson Coors to the two can suppliers. The figures under the "Actual" column are the amounts shipped from the two suppliers to the off-site warehouses near breweries. All figures in Exhibit 4-6 are Canada-wide aggregate figures. The aggregate initial inventory levels at the off-site warehouses are: 14,000,000 (355 ml) cans and 1,000,000 (473 ml) cans on July 1, 2009; and 19,000,000 (355 ml) cans and 400,000 (473 ml) cans on January 1, 2010.

Exhibit 4-6 Canada-wide aggregate usage forecasts and actual shipments (in units)

	Usage Forecast	Actual Shipment	Variance
July 2009			
355 ml	52,418,820	67,043,220	14,624,400
473 ml	4,364,969	6,691,893	2,326,924
August 2009			
355 ml	47,213,380	46,651,797	(561,583)
473 ml	6,169,380	4,651,478	(1,517,902)
January 2010			
355 ml	30,685,319	22,001,991	(8,683,328)
473 ml	5,147,421	9,615,290	4,467,869

	Usage Forecast	Actual Shipment	Variance
February 2010			
355 ml	38,388,313	27,614,402	(10,773,911)
473 ml	3,131,391	2,053,025	(1,078,366)

Discussion Questions

1. In general, what are the fundamental steps that need to be considered for the risk management process? Speculate on one Molson Coors-specific aspect in this regard that the Company should pay extra attention to in managing integrated financial and operational risk.

2. Can the Company pass on increases in costs to the market price of the final product?

3. Identify one potential risk for each of the following functions in the context of how these functions are executed at Molson Coors: i) forecasting, ii) procurement and iii) transportation. Critique the process followed in each of these three functions.

4. Speculate on the possible reasons as to why there are significant positive and negative discrepancies in the usage forecast and actual shipment figures in Exhibit 4-6.

5. In light of the financial and operational information provided in Exhibits 4-4 to 4-6, critically evaluate the interfaces between financial and operational risk management techniques used at Molson Coors for the July–August 2009 and January–February 2010 time periods. Are there opportunities missed in risk management? What would you do differently to manage risks during these peak and low demand periods?

6. Suggest some actions that Molson Coors can undertake in order to manage the volatility in aluminum prices. Speculate on possible reasons why the Company hedges against aluminum prices itself rather than letting the aluminum can supplier conduct this hedging.

7. Comment on the committee structures of CRMT and FRMC. What can you suggest to improve the decision making process within these committees?

Case 5

Toyota China: Matching Supply with Demand[1]

Xiaoying Liang[†], Lijun Ma[‡], and Houmin Yan[°]

"For years, operations managers have recognized that the matching of supply and demand is one of their most challenging problems."[2]

—*Gabriel Bitran, MIT Sloan School of Management*

Five eager faces looked up at Mr. Johnson Zhang, regional sales manager for Central China of GAC Toyota Motor Co., Ltd. (GAC-Toyota), as he entered the conference room. It was May 2009, and the meeting had been scheduled to discuss the quota allocation of a newly released Toyota SUV model, the Highlander, which had been in great demand since its launch. The five attendees were major sales

[1] Dr. Xiaoying Liang, Dr. Lijun Ma and Dr. Houmin Yan prepared this case for class discussions. It is not intended to serve as endorsement, sources of primary data or illustrations of effective or ineffective management. Preparation of this case was supported in part by RGC Competitive Earmarked Research Grants 4187/09 and 4183/11, and by NSFC grants 71001073 and 71271182. The authors would like to thank Mr. Johnson Qiang, Toyota's regional manager for central China, for his detailed introduction to the Chinese automobile market and permission to access data from dealerships.

[†] City University of Hong Kong, Hong Kong; xiliang@cityu.edu.hk

[‡] Shenzhen University, Shenzhen, Guangdong, China; ljma@szu.edu.cn

[°] City University of Hong Kong, Hong Kong; Houmin.Yan@cityu.edu.hk

[2] "Matching Supply with Demand: An Introduction to Operations Management," 3rd ed., Gerard Cachon and Christian Terwiesch, McGraw-Hill/Irwin, 2012.

representatives of GAC-Toyota in Central China, all with huge bookings from their dealers to fulfill. They each had the sole objective of getting the largest quota possible from the headquarters. However, as the huge popularity of the Highlander far exceeded the prelaunch forecast, some dealers would inevitably be disappointed due to the production capacity constraint. What could Mr. Zhang do to ensure that as many dealers and their customers as possible were satisfied? Would the supply shortage and long delivery time mean the loss of customers to competitors? Well aware of the increasingly intensive competition in China's automobile market, Zhang knew he did not have much time left to figure out the solution to the problem.

The Automobile Industry in China

Global Overview

The automobile industry is one of the most important economic sectors in the world. In 2010, over 77 million vehicles were produced worldwide.[3] China, Japan, and the United States are the top three producing countries, accounting for 46% of total production. Based on the total number of vehicles produced in 2005, then 66 million, it was estimated that the global turnover of the automobile industry was equivalent to €1.9 trillion. The industry also created over 8 million jobs directly, representing more than 5% of the world's total manufacturing employment, and about five times more indirectly. According to a study of 26 countries conducted by the International Organization of Motor Vehicle Manufacturers (OICA), vehicle manufacturing and usage contributed more than €433 billion to government revenues.[4]

---

[3] "World motor vehicle production by country and type," OICA correspondents survey, http://oica.net/wp-content/uploads/all-vehicles-2010.pdf, 2011

[4] "The world's automotive industry," OICA, http://oica.net/wp-content/uploads/2007/06/oica-depliant-final.pdf, 2006.

Historically, the automobile industry has accounted for about 3% of the gross domestic product (GDP) in the United States,[5] while in China its contribution has increased from 1.50% of GDP in 2005 to 2.61% in 2010.[6]

The top-10 manufacturers accounted for more than 66% of the total motor vehicle production of 77.8 million units in 2010. The Japan-based Toyota Motor Corporation was the leader in global motor vehicle production, with a total of 8,557,351 vehicles, followed by the U.S.-based General Motors Company (GM) at 8,476,192.[7]

History of the Automobile Industry in China

The automobile industry in China has a history dating back to the first Five-Year Plan (1953–1957). In 1953, First Automobile Works (FAW) was founded in Changchun, the capital of Jilin province. In 1956, it produced its first product, a 4-ton commercial truck under the brand name *"Jie Fang,"* which means "liberalization." In the more than five decades since then, and especially since 1978 when China began its economic reform (also known as "reform and opening up"), China's automobile industry has gradually grown into a pillar of the economy and has become one of the most important in the world (see Figure 5-1). Between 2000 and 2011, it recorded an average annual growth rate of 22%, much higher than the average global growth rate during the same period. Only in 2008 did China post a single-digit growth rate, due to the global financial crisis. In 2009, China surpassed Japan to become the No.1 automobile producing country and also topped the United States to become the largest automobile market in the world. Nonetheless, the car penetration rate (CPR, measured

---

[5] "Contribution of the automotive industry to the economics of all 50 states and the United States," Center for Automotive Research, Ann Arbor, MI., 2010.

[6] "The Overall Development Situation and Trend of the Automotive Industry in China," Yang Dong, Shanghai Auto, Vol. 6, 2011.

[7] "World ranking of manufacturers 2010," OICA, http://oica.net/wp-content/uploads/ranking-2010.pdf, 2011.

as cars per thousand people) is still low compared with developed countries (see Table 5-1), which shows huge potential in this market.

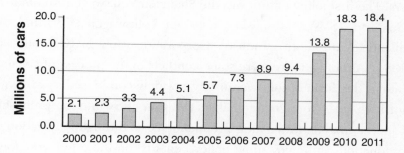

Figure 5-1 China's vehicle production from 2000–2011.

Source: Printed with permission of the China Association of Automobile Manufacturers (CAAM).

Table 5-1 List of Major Countries by Number of Vehicles per Thousand People

No.	Country	CPR	Data Year
1	China	83	2011
2	Japan	589	2009
3	U.S.	812	2010
4	Germany	634	2008
5	South Korea	379	2011
6	Brazil	259	2011
7	France	575	2007
8	Spain	608	2008
9	India	18	2009
10	Mexico	276	2009

Source: World Bank

Originally, China's automobile manufacturers were all state-owned, either by the central government, such as FAW and Second Automobile Works (Dongfeng Motors), or by local governments, such as the Beijing Automotive Industry Corporation (BAIC), Chang'an Auto, Guangzhou Auto, and Fujian Auto. With the beginning of the economic reform in 1978, the government started to allow the operation of private automakers and joint ventures with foreign automakers. The first joint venture was the Shanghai Volkswagen Automotive Co., Ltd. (SVW), established between Volkswagen (VW) and the Shanghai Automotive Industry Corporation (now SAIC Motor) in 1984.[8] Although there were some conflicts in the process of collaboration, this form of joint venture was widely considered a "win-win" for both sides: foreign automakers can gain entry into the promising Chinese market, lower their production costs through localization of production, and enjoy some policy benefits; on the other side, local manufacturers could gain access to the advanced technology, management and marketing expertise of the foreign partner and capitalize on established foreign brands. In 2011, the top-10 manufacturers sold 6,472,200 cars, accounting for 64% of total car sales (see Table 5-2). Eight of them were joint ventures.

Table 5-2 Top 10 Manufacturers by Sales in 2011

Rank	Manufacturer	Volume (10,000)
1	SAIC-GM	111.87
2	SAIC-VW	100.54
3	FAW-VW	97.63
4	Dongfeng-Nissan	66.54
5	BAIC-Hyundai	58.56
6	Chery	46.88
7	Geely	43.28
8	Changan-Ford	41.54
9	Dongfeng-Peaguot	40.41
10	FAW-Toyota	39.97
Total		**647.22**

Source: Printed with permission of the China Association of Automobile Manufacturers (CAAM).

---

[8] "Volkswagen Group China," Wikipedia, http://en.wikipedia.org/wiki/Volkswagen_Group_China.

Toyota in China

Established by Kiichiro Toyoda in 1937 as a spinoff from his father's company, Toyota Industries, Toyota Motor Corporation has grown into one of the world's leading automobile manufacturers. In 2010, Toyota produced 8,557,351 vehicles and achieved worldwide consolidated sales of 8.4 million vehicles under the Toyota, Lexus, Daihatsu, and Hino brands, ranking first in the world according to both measures.[9]

In 1959, Toyota opened its first plant outside Japan in Brazil, and it has maintained a philosophy of localizing both the production and the design of its products ever since.[10] It soon became a leader in customer satisfaction and embodied the success of Japanese automobile manufacturers in the global market. The Toyota Production System (TPS), built on the two pillars of "Just-In-Time" production and "Jidoka,"[11] is well known for its ability to reduce production costs and lead time, eliminate defects, and improve the overall quality of its products.

Although Toyota began exporting cars (starting with the Crown sedan) to China in early 1964, localization of production got off to a late start compared with Toyota's rivals. For each model, Toyota's usual strategy is to import first and then to decide whether to localize production depending on the model's popularity. Its first joint venture, the Tianjin Toyota Motor Engine Co., Ltd., only started operating in 1988. In the same year, Sichuan FAW Toyota Motor Co., Ltd. was founded, and it produced the first locally produced Toyota vehicle, the Coaster bus, 14 years later than its main competitor, the

---

[9] "World ranking of manufacturers 2010," OICA, http://oica.net/wp-content/uploads/ranking-2010.pdf, 2011.

[10] "Toyota history: Corporate and automotive," Toyoland.com, http://www.toyoland.com/history.html.

[11] "Toyota Production System Terms," Toyota Georgetown, http://toyotageorgetown.com/terms.asp.

VW group. Since then, Toyota has sped up its expansion. In 2003, the Sichuan plant began to produce the Land Cruiser Prado.[12] In 2004, Toyota established a joint venture, GAC Toyota Motor Co., Ltd., with the Guangzhou Automobile (GAC) Group to produce the Camry in Guangzhou, the capital of Guangdong province and the largest regional automobile market in China. In 2009, Toyota decided to localize the production of its popular SUV model, the Highlander, in GAC-Toyota. By 2011, Toyota had localized the production of 19 sub-brands in China. Its market share in China was 5%, compared with 18% in the United States and more than 40% in Japan.[13] Despite the low market share, the Toyota Camry, RAV4, and Highlander are among China's top sellers in the higher price ranges, generating higher profits.

4S Stores

Like other foreign automakers, the majority of Toyota's marketing, distribution, and sales operations in China are conducted by its joint venture dealerships. The dealerships are called 4S stores: Sales, Spare parts, Service and Surveys. They are designed to provide integrated services to customers. According to the statistics of the Ministry of Commerce and China Automobile Dealers Association (CADA), in 2010 there were 15,000 4S stores in China, of which 1,700 had been newly added in that year alone, and the number is expected to reach 30,000 by 2015.

Although the spike in the number of 4S stores represents the fast expansion of sales networks, profits are shrinking. When 4S stores first arrived in China around a decade ago, they had a very high rate of return on investment. Although a 4S store required an initial investment of around 20 million RMB to launch, it was fairly common for

---

[12] "FAW Toyota history," FAW, http://www.faw.com/international/toyota_history. jsp?TM=FAW-Toyota.

[13] "Sorry, Toyota: GM is winning China", Michael Brush, MSN Money, 2011.

this investment to be made back within two years or even within three months.[14] The profit mainly came from new-vehicle sales. However, as competition intensified and the profit became slimmer, 4S stores began to struggle to make a profit. J.D. Power and Associates recently conducted a survey of 1,605 4S dealerships in China, comprising 38 brands in 59 cities, which indicated that the percentage of dealers reporting a profit in 2011 fell to 63 percent, compared with 81 percent a year ago, and 20 percent of dealers reported that they had lost money on their operations—up from 9 percent in the previous year. On average, dealerships in China currently derive 40 percent of their profits from new-vehicle sales, a significantly higher proportion than in mature markets. It is expected that dealerships will gain greater profits from vehicle financing, used-vehicle sales and servicing and parts as the market continues to evolve.[15]

Toyota China's Production Planning and Demand Management

Demand Forecasting and Production Planning

To make the best use of Toyota's highly efficient TPS production system, the real challenge is in accurately forecasting demand and planning production accordingly. This is extremely difficult in a fast-growing market, such as China, particularly for newly introduced models with no historical sales data. Toyota China holds a sales convention at the end of each fiscal year, which gathers the major sales representatives from all over China. One important mission of this convention is to collect the dealers' replenishment plans for the coming year, which should include both the total quantity and the

---

[14] "China's 4S car dealerships hit the skids," CKGSB Knowledge, Sep 11, 2011.

[15] "J.D. Power and Associates reports: Amidst Beijing auto show product celebration, the industry is experiencing declining dealership profits," PRNewswire, Apr 22, 2012.

detailed numbers for specific models and configurations for each month. Toyota then determines the yearly quota for each dealer based on the numbers submitted and the consolidated sales for the past year. The general production plan for the coming year is arranged correspondingly. In implementation, Toyota China can make adjustments to the general plan according to the realized sales.

Toyota China has adopted a multi-level management structure. Regional sales managers are responsible for all of the sub-regions within their regions. They determine an overall quota and then allocate the quota among the sub-region dealers. When the demand is realized, they can also arrange transshipments between sub-regions if necessary. This centralized management helps to achieve coordination among dealers. Dealers also enjoy a certain amount of flexibility in demand realization. They must inform Toyota China of their replenishment quantities three months in advance, and they can then adjust the quantity by up to 10% until two months before the delivery. They can specify the colors up to one month before the delivery.

However, although adjustable production and replenishment can help to alleviate the risks of overstock and shortage, the effect is still limited. For instance, the Highlander, a popular SUV model, began local production in China in early 2009. During our field study of GAC-Toyota, we found that in the first half of 2009, the number of Highlanders ordered by dealerships was 60% higher and the actual realized demand was 90% higher than Toyota China's projected demand. Because some core parts were imported and the capacity was constrained, Toyota China was unable to increase its production in time to fully satisfy the surging demand. Toyota China responded to the supply shortage by expanding production of the Highlander in September 2009, although there was already a long lag. The inevitable result was a prolonged delay in the delivery time to customers, up to three to six months compared with the usual one month or less.

Customer Management at 4S Stores

According to Toyota, by the end of 2009, it had around 650 4S stores in China, which form the distribution network for its vehicles. Currently, 4S stores generate most of their profits from new-vehicle sales.

Figure 5-2 illustrates the normal operation of a 4S store. The evolutionary Customer Relationship Building (e-CRB) system is an in-house customer relationship management system used by Toyota China's dealerships. Its two key components are the intelligent Customer Relationship Optimization Program (i-CROP) and the Total Arranging and Cultivating (TACT) system. The i-CROP is responsible for the management of customer information, and the TACT is the interface between dealerships and Toyota China. The customer tracking block classifies customers into four classes (A, B, C, and D) according to their intended purchasing times. This classification is usually based on a subjective estimation by 4S store clerks. A detailed description of the classification system is provided in Table 5-3. Once orders are placed, dealerships use the TACT system to track them until final delivery. According to the 2-month data that we collected from a typical 4S store, about 20% of the in-store customers who purchased the Highlander belonged to Classes A and B, and the rest were split evenly between Classes C and D.

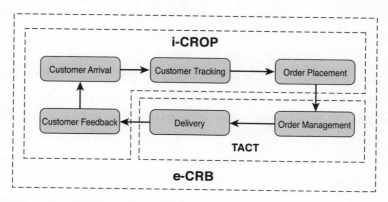

Figure 5-2 Business process at a 4S store

Source: Field study at GAC-Toyota.

Table 5-3 Customer Categories

Class	Predicted Purchasing Time
A	1 week
B	Between 1 week and 1 month
C	Between 1 month and 3 months
D	> 3 months

Source: Field study at GAC-Toyota

The interests of Toyota China and its 4S stores are not always perfectly aligned. For example, Toyota China would like to maintain central control over decisions such as pricing, bundling sales, and cross-regional transshipments to avoid image damage and malicious competition, whereas 4S stores would like to enjoy more flexibility. This interest misalignment is particularly prominent when a supply-demand imbalance emerges. As Toyota China cannot fully address the supply shortage by increasing its production, the dealerships need to make best use of their limited inventories to retain customers. In the case of a supply-demand imbalance, dealerships switch to a market segmentation strategy characterized by price and delivery-time differentiation. Two purchasing options are offered to customers: spot and consignment. By choosing the spot option, customers get their desired cars immediately from the on-hand inventory, but either need to pay a higher price, buy a bundled insurance product or upgrade the configuration. By choosing the consignment option, customers pay the manufacturer's suggested retail price (MSRP) and are put on a waiting list, with delivery usually taking two-three months. However, because the "mark-up" associated with the spot option is often determined by dealerships themselves according to the prevailing supply-demand condition, it varies from time to time and from place to place. This has caused a lot of controversy among customers, with some of them even blaming Toyota China for intentionally creating a shortage to rip off customers.

As the regional manager, Mr. Zhang needs to coordinate the actions of his dealers. However, he first needs to answer the following questions. Will the coordination of individual dealers be beneficial, for example, in terms of demand and inventory management? If so, how should Toyota China take advantage of the coordination? Regarding the customer segmentation strategy adopted by dealers, what dimensions, other than price, does the strategy explore? In addition to the profit improvement from refined customer segmentation, what other benefits can dealers and Toyota China obtain by utilizing the customer information collected using the strategy?

Case 6

Cisco Systems, Inc.: Supply Chain Risk Management[1]

María Jesús Sáenz[†] **and Elena Revilla**[‡]

"In an increasingly networked world, supply chain risk management is top of mind in global organizations as well as key differentiator for leading value chain organizations."

—John Chambers, Chairman and CEO, Cisco Systems

James Steele, program director for supply chain risk management at Cisco Systems, Inc. (Cisco), woke up one morning in March 2011 with an urgent phone call from one of the risk managers and member of one of Cisco's supply chain risk management teams based in San Jose, California. A warning system related to weather monitoring systems alerted of a high probability of suffering a high magnitude earthquake on the east coast of Japan. Once James arrived at his office, the very first activity was to open the map of key locations of facilities for Cisco's electronics components in Japan and the company's Japanese-sourced suppliers.

The threat came true, and, after a 9.0 magnitude earthquake, Japan suffered a tsunami, which caused severe damage, power failures, and meltdowns at nuclear facilities. It has been one of the largest disruptions to global supply chains in modern history.

[1] The authors prepared this case as a basis for class discussion.

[†] MIT-Zaragoza International Logistics Program, Zaragoza, Spain; mjsaenz@zlc.edu.es

[‡] IE Business School, Madrid, Spain; Elena.Revilla@ie.edu

Alarmed, James got in contact with John Chambers, Chairman and CEO at Cisco Systems, informing him that the incident management protocol for a "Black Swan" or "force majeure" event had been activated within the Global Business Operations and Supply Chain Operations Departments, at the highest degree. Steele informed, "Cisco has about 250 tier-1 suppliers in Japan, many of them the sole source for high-level engineering components." John Chambers expressed, "The main purpose now is to guarantee continuity of supply; the lowest impact in operations as well as financial performance." Steele mobilized a team of 10 risk managers to better understand the potential impact of such a disruption. The following morning, Cisco had assembled a 100-person "war room" to figure out which orders were and could be affected, accounting for all of the major suppliers within Japan, at least all the tier ones.[2]

The Company

Cisco, whose name comes from the city name, San Francisco, was founded in 1984 by Leonard Bosack and Sandy Lerner, a couple working in IT at Stanford University. Cisco is the kind of success story that is told often about Silicon Valley—the idea nurtured at home that reaches world-class level.[3]

Cisco started in the market with routers and switches. Fifteen years after foundation, routers and switches were still core products for the company, but they had expanded to include solutions based on their initial products, such as Internet Protocol (IP) communication, call center systems, Tele-Presence, and many other communication-related solutions. Cisco solutions can serve from small businesses with minor requirements, such as video-conferencing or Wi-Fi access points, to big corporations requiring Data Center and Cloud Services.[4]

Cisco started with an initial funding from the venture capital firm Sequoia Capital and went public 6 years later. That made a market

---

[2] Risky business at Cisco Systems, Robert J. Bowman. Supply Chain Brain, November 15, 2011.

[3] Cisco Systems Inc.: Collaborating on New Product Introduction, Maria Shao and Hau Lee. Case GS-66 Stanford Graduate School of Business, June 2009.

[4] Cisco Systems, Inc. 2011 Annual Report.

capitalization of U.S. $224 million, enough to be listed on the Nasdaq stock exchange. In late March 2000, Cisco had become the most valuable company in the world, with a market capitalization of more than US$500 billion. Eleven years later, in November 2011, a market capital of about U.S. $94 billion still qualifies Cisco as one of the most valuable companies.[5]

A key moment for Cisco's development was the growth of the Internet in the second half of the 90s and the big change it implied in the telecom market. Once the IP protocol was almost universally adopted, the importance of multi-protocol routing declined, and Cisco managed to become a vital supplier in a market whose growth rate was explosive. To keep up with the growth of the Internet, Cisco gained access to the required new technologies through acquisitions and partnerships. Cisco maintained its own R&D activity not only by developing new technologies and products but also by acquiring other companies, making agreements for joint-development projects, and selling products coming from other companies. This growth model was almost unique in the high-tech world. At that time, distrust among competitors and a common idea that a company looking outside for technological aid was showing its weakness prevented many companies from behaving like Cisco. Therefore, leveraging markets as well as technology was not common practice in the Industry. As a result, recently, Cisco's acquisition policy has scooped up some 160 companies over the last 10 years. "In any year, there are at least 10 companies we are trying to integrate," said Angel Mendez, Cisco's senior vice president of worldwide manufacturing.[6]

Cisco's Supply Chain

Cisco's supply chain has been evolving to a prevalence of outsourcing and globalization. Twenty years ago, original equipment manufacturers in the electronics industry were more vertically

[5] Chron 200 Market capitalization, Fost D. San Francisco Chronicle, May 5, 2006.

[6] Cisco Systems Inc.: Collaborating on New Product Introduction, Maria Shao and Hau Lee. Case GS-66 Stanford Graduate School of Business, June 2009; Cisco Systems Inc.: Acquisition Integration for Manufacturing, Nicole Tempest, Charles A. Holloway and Steven C. Wheelwright. Case OIT-26 Stanford Graduate School of Business, February 2004.

integrated, mostly performing their own manufacturing. However, nowadays, Cisco relies on contract manufacturers for almost all of its manufacturing needs, reaching more than 95% of their 12,000 products. Cisco uses a variety of independent third-party companies to provide services related to printed-circuit board assembly, in-circuit testing, product repair, and product assembly.[7]

Arrangements with contract manufacturers are carefully crafted to guarantee quality, cost, and delivery requirements, as well as capacity, cost management, oversight of manufacturing, and conditions for use of their intellectual property. Cisco has not entered into significant long-term contracts with any manufacturing service provider. The company generally has the option to renew arrangements on an as-needed basis, formalized mainly in yearly agreements. These arrangements generally do not commit Cisco to purchase any particular quantities, beyond certain amounts covered by orders or forecasts covering discrete periods of time defined by default as less than one year.

Additionally, the Cisco outsourced model became more sophisticated when it came time to internalize the challenges of Cisco's strategy of using acquisitions and partnerships in order to expand. Most of these companies had their own supply and manufacturing facilities around the globe, whose products should be delivered through Cisco's existing distribution channels. However, the main explicit acquisition criteria for deciding in which companies to integrate included not only achievement of short-term and long-term gains along with sharing similar understanding of the market or similar culture with Cisco, but also having a similar risk-taking style.[8]

This manufacturing model was explained by Carl Redfield, Cisco vice president of manufacturing and logistics at Cisco, in these terms: "Cisco wants to add value by managing the supply chain and focusing

---

[7] Cisco Systems Inc.: Collaborating on New Product Introduction, Maria Shao and Hau Lee. Case GS-66 Stanford Graduate School of Business, June 2009.

[8] Cisco Systems, Inc. 2011 Annual Report; Cisco Systems Inc.: Acquisition Integration for Manufacturing, Nicole Tempest, Charles A. Holloway and Steven C. Wheelwright. Case OIT-26 Stanford Graduate School of Business, February 2004.

on product design and development." Exhibit 6-1 summarizes Cisco's current value chain challenges, which are the main drivers for Cisco's current business model.[9]

| Outsourced Production Model | Wide Range of Products and Global Mfg Model | Products Are Primarily Configured to Order | Wide Range of Customers with Multiple Needs | Constant Acquisition Integration |

Exhibit 6-1 Cisco's value chain challenges.
Reprinted with permission of Cisco Systems, Inc.

Complementarily to outsourcing and globalization, Cisco tried to boost efficiency. In early 2006, Cisco formally introduced the orientation toward just-in-time manufacturing based on a "pull model" known as Cisco Lean.[10] The goal of this model was to convert Cisco and its extended supply chain into a system in which the product is configured to order and it is finally built only after a customer has actually ordered it, using mostly standard components.[11] Lean orientation combined with a great level of outsourcing introduced a paradigm shift within Cisco's supply chain, moving from the traditional view to the holistic end-to-end view supply chain management (see Exhibit 6-2).

---

[9] Cisco Systems Inc.: Acquisition Integration for Manufacturing, Nicole Tempest, Charles A. Holloway and Steven C. Wheelwright. Case OIT-26 Stanford Graduate School of Business, February 2004; Balance Sourcing, Laseter, T. San Francisco:Jossey-Bass, 1998.

[10] Cisco Systems Inc.: Collaborating on New Product Introduction, Maria Shao and Hau Lee. Case GS-66 Stanford Graduate School of Business, June 2009.

[11] Ibid.

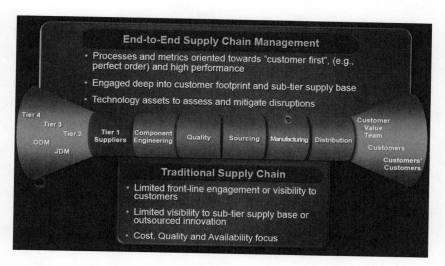

Exhibit 6-2 Holistic Cisco's supply chain.
Reprinted with permission of Cisco Systems, Inc.

In order to reach efficiency, the lean model required Cisco to have as few suppliers as possible. However, contingencies in a tense supply chain may need some levels of redundancy in sourcing to assure availability. But while reducing the base of suppliers, Cisco had to face several additional challenges from the lean model. They needed to reduce inventories while hoping for shorter and more predictable lead times across an easier to control and synchronize extended supply chain. A lean approach also encourages component standardization. It faces product differentiation demanded by the market through different configuration of these standard components.[12]

Cisco's key supply-chain decision makers understand supply chains as dynamic systems that need to be continuously adapted to the current context in which they are operating. "Nowadays, activities are much more intertwined from an operational as well as financial point of view. We are driving an adaptive supply chain in a very large

---

[12] Cisco Systems Inc.: Collaborating on New Product Introduction, Maria Shao and Hau Lee. Case GS-66 Stanford Graduate School of Business, June 2009; Cisco addresses supply chain risk management, Miklovic D. and Witty R. Case Study ID G00206060 Gartner Industry Research, September 17, 2010.

outsourced model across a very large spectrum of products and geographies. This is somewhat unique," expressed Mendez.[13] This way of managing the supply chain placed Cisco's supply chain in the sixth position in the prestigious Gartner's Supply Chain Top 25 in the 2011 ranking.

The consequences of this pattern of changes toward a disperse network have been quite relevant. Cisco's supply chains have become extensively stretched to significant levels of supply chain dependence on globalization. The opportunities provided by the benefits of outsourcing together with globalization have been the main driver for supply chains to fully focus on seeking efficiency. But, this efficiency is not free, and now Cisco's supply chains are facing higher vulnerabilities.

Supply Chain Risks

Today's electronics supply chains face risks from many factors including political upheavals, regulatory compliance mandates, increasing economic uncertainty, rapid changes in technology, demanding customer expectations, capacity constraints, the effects of globalization and natural disasters. "These threats may come from both internal and external sources of risk," explained Steele. The first one, related to contingencies inside the supply chain, has to do with how the supply chain is looking for efficiency.[14] Some actions can be taken by the supply chain manager to prevent and control these internal forces—for instance, process orientation action, prevention, monitoring supply chain key nodes, and visibility.

External sources of risk arise from the organization's exposure to the external environment (i.e., natural hazards, economic sources, or market sources of risk). Compared to the first set of sources, external risks are mainly influenced by globalization where the company has imperfect and incomplete control over the external environment.

---

[13] Cisco Systems Inc.: Collaborating on New Product Introduction, Maria Shao and Hau Lee. Case GS-66 Stanford Graduate School of Business, June 2009.

[14] What is the right disruption management for your supply chain?, Revilla E. and Saenz M.J. Forbes India, October 28, 2011.

Cisco is aware of how vulnerable its business may be due to the current world scenario affected by the global economic downturn. This environment marked by market uncertainty brings potential threats to which Cisco is paying attention, such as foreign currency exchange rates, economic weakness, adverse tax consequences, political or social unrest, and trade protection measures. Some of them may affect the firm's ability to import or export products, which is a key activity in a seamless supply chain within a global market.[15] To diminish the potential effects of external risks, the involvement of C-level executives is required as well as collaboration with stakeholders. Monitoring the market, the political scenario, the weather, the economy and other extraordinary events is also essential.[16]

At the time of the tsunami in Japan, Cisco optical server routers were configured to order, having a disperse network of diverse tier and sub-tier suppliers coming from different world regions (see Exhibit 6-3). Suppliers of raw materials such as resins, direct tier-1 suppliers of components such as complex optical line cards, and intermediate sub-tier suppliers are all potential single points of failure that could start a chain of harmful reactions. "Now we don't have tons of inventories when a disruption surprises to us. But our customers for optical service routers desire products with low lead times and high responsiveness levels," exclaimed Steele, regarding the restriction of inventories within a global network base of suppliers.[17]

---

[15] Cisco Systems, Inc. 2011 Annual Report.

[16] What is the right disruption management for your supply chain? Revilla E. and Saenz M.J. Forbes India, October 28, 2011.

[17] Cisco addresses supply chain risk management, Miklovic D. and Witty R. Case Study ID G00206060 Gartner Industry Research, September 17, 2010.

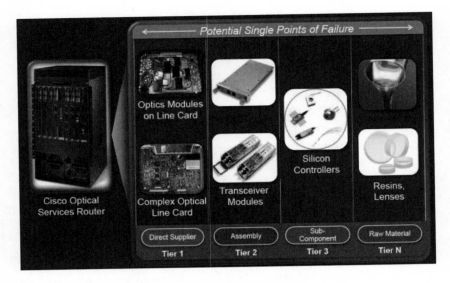

Exhibit 6-3 Cisco's optical service router supply chain.
Reprinted with permission of Cisco Systems, Inc.

When the tsunami happened, the vulnerabilities of the optical service routers supply chain became evident. In fact, one of the main raw material suppliers lost more than 50% of its capacity. A supplier for assembled components faced important problems due to its location close to the tsunami area. Transportation carriers, moving some subcomponents to this assembler, failed. Other suppliers faced product quality issues due to the lack of manpower. Additional effects occurred as a consequence of these supply chain discontinuities. For instance, manufacturing lead times became longer, and some customers, so as to ensure availability, made the same orders twice or even three times through several sales channels and later cancelled the orders upon receipt of the product. This effect, along with other factors, caused important distortions in sales prediction and subsequent inventory assessment.[18]

Furthermore, Japan's tsunami brought the importance of cultural differences to light. The supply network for this router based in Japan was culturally heterogeneous compared to Cisco's U.S.-based command, with different ways and procedures for perceiving threats as

---

[18] Cisco Systems, Inc. 2011 Annual Report.

well as risk assessment and mitigation. "Not all our suppliers and other stakeholders had the same risk specification. From Cisco's headquarters we were worried from the very first tier supplier to guarantee long term end-to-end supply. While one of the raw material suppliers promised that the tsunami was going to have very minimum effect on Cisco's components, we knew that its capacity had been extremely damaged," explained Steele. Such was the relevance of the cultural issues, that they were treated as critical in the overall supply chain risk management at Cisco.[19]

Supply Chain Risk Management

"Cisco's proactive approach and leading supply chain risk management capabilities were key to ensuring minimal impact to our customers during the Japan earthquake crisis."

—John Chambers, Chairman and CEO, Cisco Systems

Cisco started to understand the importance of managing risks for ensuring supply chain performance when they started to deploy the business continuity planning (BCP) in 2004 (see Exhibit 6-4). Cisco's BCP program aimed to implement actions for risk mitigation, but these efforts were mainly *reactive* in nature since they were basically managing a crisis when it occurred. It indentified critical business processes, people, and systems related to their supply chain network. They built a BCP dashboard with data for helping the crisis management team to determine the impact of any disruption. This allowed them to achieve a certain level of mitigation. They categorized the potential disruption based on the result of multiplying the risk of the incident by the financial impact. [20]However, the reality showed that this reactive incident-based risk approach proved to be of minimal use since "we were always guessing wrong."[21] It was just useful into "scaring the business" into providing funds to continue developing

---

[19] Cisco strives to identify and mitigate risk in its supply chain. Supply Chain News, Digi-key, PurchasingPro, October 6, 2011).

[20] Innovating through Supply Chain Risk Management, John O'Connor. 94th Annual International Supply Management Conference, May 2009.

[21] Based on an interview with James Steele, Program Director, Supply Chain Risk Management, Cisco, March 2012.

the supply chain risk management program, as well as establishing the basis for starting the maturation of a very advanced approach.[22]

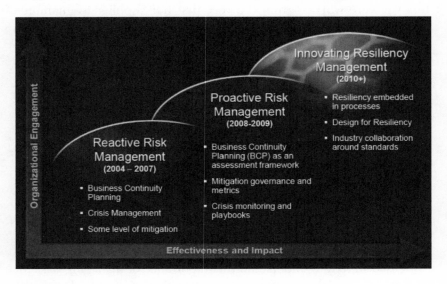

Exhibit 6-4 Evolution of supply chain risk management at Cisco.
Reprinted with permission of Cisco Systems, Inc.

Hurricane Katrina in 2005 instigated the challenge of taking a step forward from Cisco's reactive approach for managing supply chain risks. In the aftermath of the hurricane, Cisco released more than $1 billion into the distribution channel to aid telecom infrastructure recovery. But Cisco was not able to assess where the product was or the impact on the company from a financial point of view.[23] "We needed a more *proactive* management, complementary to the previous one that would allow us to use the business continuity planning as an assessment tool," Steele remarked.[24]

As a result of these reflections, the company developed beforehand response playbooks that listed strategies and responsibilities for

[22] Cisco addresses supply chain risk management, Miklovic D. and Witty R. Case Study ID G00206060 Gartner Industry Research, September 17, 2010.

[23] Cisco addresses supply chain risk management, Miklovic D. and Witty R. Case Study ID G00206060 Gartner Industry Research, September 17, 2010.

[24] Based on an interview with James Steele, Program Director, Supply Chain Risk Management, Cisco, March 2012.

taking action, specified by the type of catastrophe and disruption type as well as the specific region and anticipated duration. That provided a framework for organizing an incident response team that monitored the crisis and the subsequent events that might disrupt Cisco's supply chain. Such assessment is particularly relevant for highly engineered components in electronic commodities, such as Cisco's. The more advanced business continuity plan helped to identify supply chain nodes and assess the Time to Recovery for each node. Time to Recovery is based on the longest recovery time for any critical capability within a node, and it measures the time required to restore 100% output at that node when a disruption occurs.[25] It can be measured against multiple factors at the component level as well as at the manufacturing and test levels.[26] In reference to this, within 48 hours after the 2008 Chengdu earthquake in China, Cisco was able to conduct a full impact analysis, assessing the time needed to restore the network. This fostered complete visibility into the supplier footprint in the area and the initiation of a crisis survey targeted at the suppliers' emergency contacts.[27]

Specifically, for the financial crisis at the end of 2008, Cisco deployed a financial risk assessment with the main purpose of categorizing single-sourced parts suppliers that could have an important impact on Cisco's economy in green, yellow, and red. When five of the red suppliers ended up in bankruptcy, Cisco had already instituted them as "last-time buys" or "second sourcing." The result of this pre-assessment was conducted with no final effect for the operations run by Cisco from such a suppliers' disruption.[28]

At this point, Cisco was aware of how dynamic its supply chains were, driven by globalization, outsourcing, and efficiency. This dynamic required establishing coherence between the supply chain evolution and the supply chain risk management evolution. The way

[25] Innovating through Supply Chain Risk Management, John O'Connor. 94th Annual International Supply Management Conference, May 2009.

[26] Based on an interview with James Steele, Program Director, Supply Chain Risk Management, Cisco, March 2012.

[27] De-Risking the Supply Chain. U.S. Resilience Project, August 8, 2011.

[28] Cisco addresses supply chain risk management, Miklovic D. and Witty R. Case Study ID G00206060 Gartner Industry Research, September 17, 2010; De-Risking the Supply Chain. U.S. Resilience Project, August 8, 2011.

they configured and designed a supply chain introduced external and internal risks, and therefore an additional effort for better managing risks was needed from the very start of the supply chain design. This was in 2010 when they started a new stage in this challenge by adding what they called the *"resiliency* management innovation" to the combination of the reactive-proactive approaches already established at previous stages (refer to Exhibit 6-4).

"We realized the real potential to assess the ability of our supply chains so as to return to the original operational state after being disturbed by external or internal factors. However, this was not proactive enough," mentioned Steele when referring to how they matured their understanding about resiliency by taking this next step into the evolution of their supply chain risk management.[29] This meant integrating the resiliency concept as part of product innovation as well as supply chain process innovation and integrating the notion of "design for resiliency." Cisco upper management was aware that this implied a significant cultural change internally at Cisco for the organizational engagement required, as well as externally in its expanded and diverse value chain. This new stage implied to de-risk the products and supply chains (their nodes, suppliers, equipment, manufacturing, and logistics) beforehand, in order to be prepared when an important incident could damage them.[30]

For product de-risk, Cisco selects alternative components in the bill of materials, uses already existing components and component risk buffers, qualifies additional manufacturing sites, and specifies alternative test procedures. This is when these products can enter production with low risk indexes. All these efforts for de-risking a product should be combined with efficient demand forecasting tools for avoiding product shortages. As a result, Cisco estimates that a de-risked product can save approximately U.S.$1 million.[31]

For supply chain de-risk, Cisco integrates risk awareness while innovating the supply chains. It consists of proactive efforts in the

---

[29] Based on an interview with James Steele, Program Director, Supply Chain Risk Management, Cisco, March 2012.

[30] Cisco addresses supply chain risk management, Miklovic D. and Witty R. Case Study ID G00206060 Gartner Industry Research, September 17, 2010.

[31] Ibid.

design and execution of the supply chain, in terms of processes, manufacturing sites, transportation routes, and external services, with the main purpose of reducing post-disaster recovery. The supply chain resiliency team works closely with manufacturing operations, and logistics and transportation service providers and partners to identify network nodes that are out of risk qualification tolerances as well as to develop resiliency plans.[32]

In March 2011, James Steele spoke about Cisco's supply chain risk management evolution: "In the past, supply chain operations were 'cared-about' only when things went wrong. The focus was not on increasing the business, but on keeping the trains running on time. Over the past 15 years, there has been a sea of change in supply chain management. It has become a strategic capability for many companies, and it continues to get the resources, visibility, and focus needed to manage as a platform for growth. For Cisco, this change has meant an increase in risk intelligence and agility on supply chain resiliency capabilities, which are a key element in this evolution."[33]

Seventy Days After the Japanese Tsunami

After launching an intensive 70-day effort at Cisco to mitigate the impacts of the earthquake in Japan, John Chambers, CEO, summoned a debriefing meeting with James Steele, responsible for the supply chain risk management team, John O'Connor, senior director of business transformation, and key members of the supply chain management team for Cisco optical service routers. All met together in order to analyze the final impact of the devastating incident.

Steele started the meeting with a presentation highlighting the key results of Cisco's response in Japan (see Exhibit 6-5). He stated:

Our incident management protocol achieved a record time in the assessment of our network in Japan. In 12 hours, the impacts of the event in all of our suppliers, from tier-1 to raw materials, were identified [up to 300 suppliers]. We collected information about the sequence of the incident. The assessment of the sub-tier suppliers' impact on Japan's map

---

[32] De-Risking the Supply Chain. U.S. Resilience Project, August 8, 2011.

[33] Ibid.

for Cisco optical service routers at that point in time is presented [in Exhibit 6-6]. Due to the endeavor of building proactive resiliency while we designed this supply chain, we have been able to provide visibility of sub-tier risk. This has been the Achilles' heel for deploying the right mitigation actions. Thanks to the very short-term effort, customer value teams were able to be positioned to liaise with customers, assessing the customer footprints across the incident occurrence. This resulted in 118 inquiries each managed within a 24-hour response-time-window. We need to receive the final evaluation from the finance department, but we can guess now that Cisco is not going to be impacted by almost any revenue lost.

Supplier Management	• In 12 hours: all suppliers identified in impacted region
	• Assessed over 300 suppliers
	• Assessed over 7000 part numbers and assigned risk rating and mitigation plan
Customer Management	• Customer value teams positioned to liaise with customers
	• Managed 118 customer inquiries, 24-hour responses
Overall result	• Almost no revenue impact

Exhibit 6-5 Key results for Cisco Japan response.

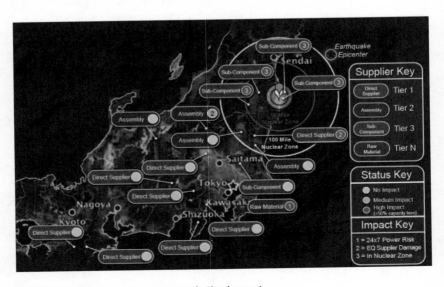

Exhibit 6-6 Mapping to assess sub-tier impact.

Reprinted with permission of Cisco Systems, Inc.

When Steele finished his excellent presentation, Cisco's CEO addressed the meeting attendees: "Being successful in the electronics industry and implementing the greatest technology is not enough, because outsourcing and globalization make the supply chain more and more vulnerable. Disruptions provide a unique opportunity to enhance our capabilities." He was aware of the important implications and impact of the efforts for mitigating the potential impact of an event of such magnitude, and he opened the discussion with the following questions for debate:

- Has the supply chain risk management implemented after the tsunami in Japan resulted in positive outcomes? What worked and what did not? Why?

- How should supply chain complexity align with supply chain risk management?

- Which comparable and comprehensive metrics have been required to assess supply chain resiliency capabilities? How can we relate these metrics to the Time-to-Recovery measurement for all capabilities?

- Can this approach to managing supply chain risks be generalized for any supply chain?

- How can the existence of the supply chain risk management efforts and its budget be justified? Does the return on investment of such efforts provide enough justification for their existence?

- What lessons can be extracted from the experience of managing these high magnitude disruptions? How might Cisco continue learning in order to improve the performance of the supply chain risk management?

Case 7

BESSI: The Importance of Coordinating Product Development with Supply Chain Planning in the Fashion Goods Industry

Maria Caridi[†], Margherita Pero[‡], and Antonella Moretto[°]

Company Background

Founded in 1958 in the southern Italian city of Naples, BESSI, as a manufacturer and retailer of fine leather handbags, opened its first store outside of Italy in Paris in 1965, followed by many others in important world cities. With its association with royalty and film stars, the BESSI brand soon became a byword for luxury. Starting from the late 1970s, BESSI extended its product range by offering a wide variety of leather goods (such as handbags, gloves, belts, suitcases), and in the 1980s–1990s, it launched the BESSI Accessories Collection, including eyeglasses, jewels, perfumes, and watches.

Over the years, design and product development have always been the core of BESSI's success, and the company has consistently striven to maintain an outstanding brand image. BESSI designs and distributes very-fashionable high-quality products. Eyeglasses and perfumes are produced and distributed by outstanding licensees. The critical success factors of BESSI include craftsmanship, quality, and made-in-Italy. Nowadays BESSI is the 50th most famous brand worldwide. The estimated value of the brand is about $7 billion. BESSI's turnover exceeds $2 million in continuous double-digit increases in recent years.

[†] Politecnico di Milano, Milan, Italy; maria.caridi@polimi.it

[‡] Politecnico di Milano, Milan, Italy; margherita.pero@polimi.it

[°] Politecnico di Milano, Milan, Italy; antonella.moretto@polimi.it

BESSI's market is divided into four macro regions: the United States, Europe, the Far East, and Japan. The turnover is equally distributed among the four macro regions. Some differences among regional demands sometimes occur due to changes in tourist flows (e.g., September 11th, avian flu).

BESSI has six distribution channels:

- **Direct Operated Stores (DOS):** More than 200 stores in 60 countries. These represent the most important channel (70%–80% of the overall turnover). Each store displays the whole collection, included the high-end items. BESSI uses DOS as a showcase for reinforcing its brand (flagship stores). Moreover, these stores allow better control of distribution (e.g., discounts, parallel markets).

- **Multi-brand stores (authorized stores):** These are located in European and U.S. cities where BESSI's DOS are not present. They collect 20%–30% of the overall turnover. In order to keep the risk of unsold goods to a minimum, multi-brand stores usually buy BESSI's low-medium price products, which have a high probability to be sold.

- **Franchising stores:** These are mono-brand stores, delegated to external partners. BESSI adopts this approach in India, Russia, the Middle East, Asia, and South Africa, where the company does not have enough local expertise or where it would be impractical to establish DOS. Each store is free to select the collection to display, although the franchisees are required to have a certain amount of items of some specific products, considered as "must have" by BESSI.

- **Duty-free shops:** These are located mainly in Far East.

- **E-commerce:** Since 2004 in the UK, 2006 in France and Germany, and 2007 in Italy, BESSI has introduced this new channel. To avoid the risk of losing control of the shopping experience, especially for high-luxury products, e-commerce is generally used to sell low- to medium-priced products. This choice is consistent with the fact that e-commerce is mainly used by young customers.

- **Outlet:** BESSI has introduced a few outlets, mainly in Italy and Europe, to sell unsold products at the end of season at a discounted price.

BESSI Leather Goods

BESSI's range of leather goods encompasses four product categories: handbags, small leather goods (such as wallets, key cases, small gadgets), belts, and suitcases. In each season, BESSI manages about 5,000 SKUs divided as follows: 35% handbags, 35% small leather goods, 15% belts, and 15% suitcases. BESSI's products are mainly *fashion products* (new products launched during a collection, with a life cycle of, at maximum, one season). As a matter of fact, only 30% of BESSI's turnover is due to carryover products (products maintained the same over time, for more than one collection). This percentage increases up to 50% in the case of some products—e.g., belts. Thus, planning and managing BESSI's operations are complex. Additionally, leather goods production is technologically underdeveloped if compared to the clothing industry or to the shoe industry. The reason is tightly connected to raw materials—leather properties are not uniform like those of industrial goods. The typical characteristics of leather, such as variations of grade, shade, or shape, as well as the presence of small scars, make the usage of automation difficult for leather goods manufacturing. The headquarters in Naples is responsible for managing operations for all leather products. All of the production phases are outsourced. In fact, BESSI relies on a network of suppliers who manufacture leather goods. All of the suppliers are Italian craftsmen, and most of them (95%) are located near BESSI. They are small companies (often family companies) whose revenues have been for years almost completely generated by their business with BESSI. Craftsmen cut and assembly (*façon*) leather for BESSI. They have a long-term relationship with BESSI, based on mutual trust. On the one hand, BESSI orders constant quantities and respects regular payments. On the other hand, craftsmen provide BESSI with products of outstanding quality, and BESSI can trust that they will deliver on time in the required quantities.

Some end products are manufactured by contract manufacturers, using materials that BESSI buys and sends to them. Some materials are directly delivered to the contract manufacturers by material suppliers, while other materials are first delivered to BESSI's warehouse in Naples and then on to the contract manufacturers. Certain other products are totally managed by end-product suppliers. In this case, suppliers buy materials, manufacturer the product, and deliver it to the BESSI warehouse in Naples (see Figure 7-1).

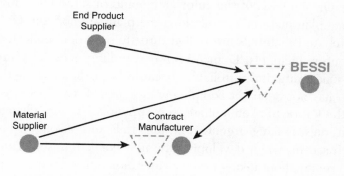

Figure 7-1 BESSI' s upstream supply chain.

On average, material purchasing lead-time is about seven weeks, whereas contract manufacturer lead-time is seven weeks further. Thus, lead-time of a product along BESSI's supply chain amounts to 14 weeks. In addition, end-product suppliers need, on average, 14 weeks for delivering end products once BESSI issues an order. The previously stated lead-times include safety lead time, so that they are reliable for planning operations activities.

The Central Role of Planning

The main activities performed by BESSI's Operations department involve planning. About 20 persons are employed as planners. Three typologies of planners are present: end-product planners, material planners, and production planners (i.e., contract manufacturer planners). End-product planners make completely autonomous decisions. Once they receive the demand forecasts from the Merchandising department (aggregate quantity for the season), each planner determines the sourcing plan of the finished goods that he/she is responsible for.

On the contrary, material planners' and production planners' decisions are expected to be tightly connected. Indeed, contract manufacturers can only manufacture end products if materials have punctually reached their shops! Unfortunately, the two plans often fail to be coordinated. In fact, material planners determine their plan starting from the provisional demand forecast provided by Merchandising. On the basis of this information and of some lot-sizing criteria, they plan material arrival along the planning horizon. Only 70% of purchased materials are pulled directly by orders collected from distribution channels. Demand forecast pulling material sourcing is often inaccurate and unreliable, because the collection for the new season has not been yet completely designed. Final decisions taken after the Fabric to Sketch phase (see Appendix 7-2) may change most of materials, making urgent some materials and obsolete some others. Moreover, when developing the new collection, some materials are chosen without taking into consideration either their actual availability in the market or the sourcing times. This may have disastrous consequences on material planning and on actual material availability.

Unlike material planners, production planners determine most of their plan on the basis of orders collected during catwalk/showroom presentations. Only a small fraction of finished goods are planned on the basis of Merchandising's forecasts. Production plans are basically pulled by orders, both in time and in volume. BESSI'S production capacity consists of the sum of the production capacities of BESSI's suppliers. Thus, production planners might increase or decrease the number of suppliers involved in each period based on the current demand level.

When planners have decided the exact quantity they want suppliers to manufacture, they insert into the information system an "order proposal" that is confirmed by material planners only if the corresponding materials are available. Materials are often not available, and so material planners have to expedite them. In these cases, if Q represents the quantity proposed by the production planners, material planners are able to confirm just a fraction of Q, e.g., $Q/2$. In order to cope with the lack of materials, over the years, production planners have over-ordered to prevent poor customer service. Instead of ordering Q, they order a higher quantity, say $2Q$. If material planners confirm $(2Q)/2$, production planners will have the desired quantity.

Unfortunately, in the cases when material planners actually confirm 2Q, the corresponding end-product quantity will be manufactured, thus creating undesired inventory. Thanks to this mechanism, the probability of end-product unavailability is lower, but the inventory level has risen too high for this make-to-order company.

Each season a new collection is launched, and the past collection soon becomes obsolete. The risk of obsolescence of end products is very high. The only way out is to sell unsold products via the outlet channel, whose selling prices are obviously lower than the regular ones. The demand forecast percentage error ranges from -20% to 40%. Forecast error varies both collection-by-collection and product-by-product. In general, a lower forecast error has been observed for carry-over products and a higher forecast error for fashion items. Figure 7-2 illustrates the probability distribution of percentage forecast error for both carry-over and fashion products.

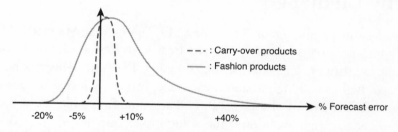

Figure 7-2 Distribution of percentage forecast error for carryover and fashion products.

By analyzing the forecast errors in detail, the probability of over- and under-estimated demand is computed, as shown in Table 7-1.

Table 7-1 Probability of Overestimating and Underestimating Demand

	Carryover	**Fashion Products**
Probability of underestimated demand (Demand>Forecast)	70%	60%
Probability of overestimated demand (Demand<Forecast)	30%	40%

For materials, thanks to the high level of commonality among materials and end products (e.g., the same metal component is assembled on the handbag, regardless of color), for certain materials the distribution of materials consumption forecast error is even more concentrated around zero than is the demand forecast error. Thus, the standard deviation of percentage forecast error is lower. Table 7-2 summarizes the range of percentage forecast error of materials, depending on the level of commonality and on the type of finished product using those materials.

Table 7-2 Range of Percentage Forecast Errors for BESSI Materials

		Commonality		
		Low	**Medium**	**High**
Finished Products	Only carry over	-5% to +10%	-3% to +5%	-2% to +2%
	Both	-10% to +20%	- 8% to +10%	- 3% to +5%
	Only fashion	-20% to +40%	-10% to +20%	-5% to +10%

The Challenge

At the beginning of 2012, BESSI's Operations Manager, Mr. Giorgi, is charged by the CEO, Mr. Brown, to find a solution to the poor logistics performance of the company. Delayed deliveries have grown higher and higher, and meanwhile, inventory levels are out of control. Mr. Brown observes, "Products are not available when they are needed. High inventories mean high costs because of our risk of product obsolescence. Late deliveries mean losing market share since customers buy the products of our competitors! What are the reasons for this? And what might be the solution?"

Mr. Brown has given Mr. Giorgi one month to find a solution. "One month is a very short time," says Mr. Giorgi. "The new information system we are implementing will not be able to provide satisfying results within one month." For a while, Mr. Giorgi has believed that the new information system will solve problems in the Operations department. He has hoped that a newly integrated "Sales and Operations Planning" system, together with a new Material Requirements Planning procedure, will definitely assure on-time delivery and low inventory. Now he is starting to believe that the information system can do very little by itself "within a month!"

Mr. Giorgi decides to charge Dr. Castelli, his assistant, to analyze the situation. "I am sure that in the business school where you have just completed your Masters degree with honors, they taught you how to manage problems of material management as complex as the one we are currently experiencing," Mr. Giorgi says. Then, he asks Dr. Castelli to analyze the main weak points of their planning process and to make some proposals in order to solve them, "In no more than two weeks!" he adds.

Collection Definition (See Appendixes 7-1 and 7-2)

Fashion Product Collection Definition

The starting point of the fashion product collection definition process is the Grid definition. The Merchandising department specifies how many different typologies of products the new collection should encompass (e.g., the collection should include three different kinds of high-end shoes). Merchandising defines the Grid on the basis of historical sales data provided by regional buyers. Meanwhile, the Research department provides another important input to the collection definition process. They continuously analyze new stylistic trends all over the world so that the department can inform the Style department about new materials, new colors, and new accessories to use within the new collection. Once the Grid is ready, the Style department starts creating the new collection. About one month later, the collection draft is shown to the Product Development department. During this meeting, about 80% of sketches are modified or rejected. For the remaining 20%, the Product Development department defines the line list, which specifies which materials to use as well as the production timetable. Once the line list is defined, prototype development starts. Prototypes are developed as soon as the Style department asks for them, according to a First-In, First-Out (FIFO) rule.

Meanwhile, the Costing department begins the evaluation of product costs with the final aim of verifying whether the selling prices of the products are consistent with the Grid. Cost evaluation is a difficult activity because it is initially defined on the basis of a sketch and not on the basis of a physical product. In particular, at first the cost evaluation is done for those items with low prices and for carryover products. Low-price products generate high volumes; therefore, cost evaluation must be accurate. For carryover items, industrial cost should always be promptly updated in the information system. Product cost becomes more accurate as the collection definition reaches its conclusion. The final cost and the price list are usually released when the final collection is presented.

After two or three weeks from the start of prototype development, the Style department previews the prototypes. At this juncture, about 50%–60% of the products are modified or rejected. About two weeks later, the Style department provides a final review of prototypes. The outcome of the review is the definition of the pre-editing collection. The pre-editing collection is further developed by the Prototype department, and after about three more weeks the "Fabric to Sketch" is defined. Fabric to Sketch is a critical point of collection definition process. The Style, Merchandising, and Product Development departments are involved in the definition of the range of colors and of fabric/leather of each product.

In the following month, in cases where internal prototype production capacity is not high enough to satisfy the peak demand for prototypes, external prototype makers are asked to make prototypes (in several combinations of color, fabric/leather). Subsequently, the Style, Merchandising, and Product Development departments finally review prototypes during the Editing phase. This review definitively defines the collection that will be presented during the catwalk and in the showroom. Notwithstanding, some further changes are possible when external prototype suppliers prepare final samples. When the final presentation (catwalk/showroom) takes place, the prototypes are definitely ready.

Carryover Collection Definition

The collection is made up of both fashion and carryover products. In recent years, the percentage of carryover products has grown. In each collection, about 80% of carryover products are lightly modified (mainly colors and materials). The carryover collection definition process is quite simple and has the aim of approving changes to carryover products. Similar to fashion products, on the basis of historical sales data, Merchandising defines the carryover products to include in the collection. After the fashion collection draft is presented and the line list is defined, Merchandise communicates to Product Development the carryover selection. Then the carryover prototypes are developed (according to an approval process that is much simpler than that of fashion products) so that they are ready for final presentation to Style and Merchandising. Both fashion and carryover products are evaluated in order to define which products actually will be presented in catwalks and showrooms.

Demand Planning Process

The demand planning process of BESSI is tightly connected with the collection definition. When analyzing the fashion product collection definition, the complexity of the process of prototype approval emerges. Prototypes are expensive, and often materials are not available when they are needed. It should be noted that production capacity of external prototype makers is high (they can handle up to 1,050 items in 10 days, which is high above the volume required by BESSI's Style department). On the contrary, material availability is a major problem. On-time physical prototypes ready for Fabric to Sketch are often few or are produced with temporary materials. As a consequence, the Style department sometimes has to make decisions only on the basis of sketches or with in-progress prototypes. This makes product and cost evaluation difficult. Unreliable and slow costing makes demand forecasting inaccurate. Moreover, the approval process becomes longer. Fabric to Sketch is subject to further revisions before final approval. A brief description of the demand forecasting process is provided next.

Demand Forecasting (See Appendix 7-3)

Demand forecasting is critical for material sourcing. During the prototype development phase, the first provisional forecast is required in order to define which materials and which quantities will be needed for the collection so that the Sourcing department can start buying them. For this reason, merchandising requests with high priority the cost of fashion items. In order to define the provisional forecast, the regional growth objectives are taken into consideration. Moreover, regional buyers evaluate the market potential for their regions, and, by comparing the potential with the items in stock, they define the "Open-to-Buy." On the basis of the Open-to-Buy, Merchandising develops the aggregate provisional forecast. According to a top-down approach, regional buyers define the forecast at item level by decomposing the aggregate provisional forecasts. Meanwhile, the Planning department defines the delivery time of the products. The final forecast is ready within the Editing phase.

Discussion Questions

1. What are the weak points of BESSI's planning process?
2. How can we help Dr. Castelli in carrying out this hard job?

Appendix 7-1 Gantt Chart of the Fashion Product Collection Definition (from Grid Definition to Start of Production)

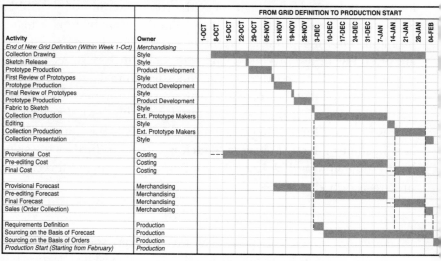

Activity	Owner	1-OCT	8-OCT	15-OCT	22-OCT	29-OCT	05-NOV	12-NOV	19-NOV	26-NOV	3-DEC	10-DEC	17-DEC	24-DEC	31-DEC	7-JAN	14-JAN	21-JAN	28-JAN	04-FEB
End of New Grid Definition (Within Week 1-Oct)	Merchandising																			
Collection Drawing	Style																			
Sketch Release	Style																			
Prototype Production	Product Development																			
First Review of Prototypes	Style																			
Prototype Production	Product Development																			
Final Review of Prototypes	Style																			
Prototype Production	Product Development																			
Fabric to Sketch	Style																			
Collection Production	Ext. Prototype Makers																			
Editing	Style																			
Collection Production	Ext. Prototype Makers																			
Collection Presentation	Style																			
Provisional Cost	Costing																			
Pre-editing Cost	Costing																			
Final Cost	Costing																			
Provisional Forecast	Merchandising																			
Pre-editing Forecast	Merchandising																			
Final Forecast	Merchandising																			
Sales (Order Collection)	Merchandising																			
Requirements Definition	Production																			
Sourcing on the Basis of Forecast	Production																			
Sourcing on the Basis of Orders	Production																			
Production Start (Starting from February)	Production																			

Appendix 7-2 Fashion Product Collection Definitions

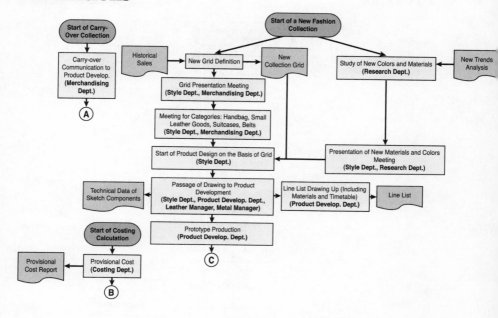

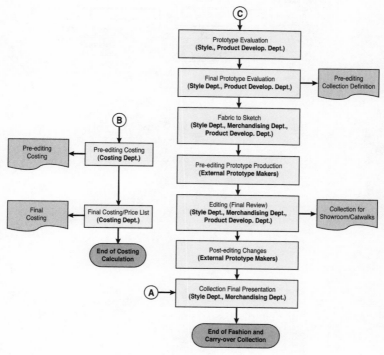

Appendix 7-3 Demand Forecasting Process

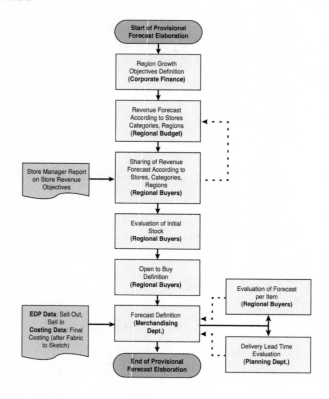

3
Supply Chain Analytics

Case 8

Queuing at eCycle Services

Janice Eliasson[†] **and Brent Snider**[‡]

The Issue

Kevin Johansson, the owner of eCycle Services, has just returned from vacation to find a $30,000 "waiting fee" invoice from the city: *"Thirty thousand dollars—are you kidding me? That was our entire profit last month—this cannot be right! I can understand the idea behind charging us while their trailers are at our facility, but $30,000 in just one month is absurd. I'm going to call the city right now and tell them they have made a massive mistake."*

Background

Located in Vancouver, British Columbia, eCycle Services is an electronic waste (e-waste) recycler that focuses on recycling cathode-ray-tubes (CRTs) that are found in older televisions and computer monitors. The owner, Kevin Johansson, started the business just over a year ago after previously working at another e-waste recycling business. After bidding and winning one of the CRT recycling contracts with the City of Vancouver, Kevin started the business by renting warehouse space and a single loading dock in an industrial area. The continued growth of CRT recycling is creating challenges for this small business working in the reverse supply chain industry.

[†] University of Calgary, Calgary, Alberta, Canada; janice.eliasson@haskayne. ucalgary.ca

[‡] University of Calgary, Calgary, Alberta, Canada; brent.snider@haskayne. ucalgary.ca

Manufacturers have switched to producing LED, LCD, and Plasma displays for televisions and computers for their performance advantages, reduced energy consumption, and supply chain benefits. The older CRT design required a large box size and weight (a 20-inch unit weighs approximately 50 pounds), creating significant supply chain costs compared to the much thinner and lighter new display technologies. Recognizing the hazardous contents of CRTs (such as lead and phosphors), most governments have legislation in place prohibiting CRTs from going to landfills. For example, the United States Environmental Protection Agency has established recycling requirements for CRTs since 2001, and the European Union's Waste Electrical and Electronic Equipment (WEEE) Directive, implemented in 2003, includes regulations for CRT disposal.

As more and more consumers replace their CRT televisions and computers, demand for CRT recycling continues to grow. Governments now provide e-waste drop-off locations, and more and more businesses now accept e-waste as well. While these locations collect the e-waste (product acquisition), it still must get transported (reverse logistics) to a recycling facility (such as eCycle) where it can be inspected, disassembled, and sorted (see Exhibit 8-1). Although some e-waste recyclers choose to simply remarket their e-waste to foreign countries for recycling, eCycle only uses reputable recyclers for their e-waste components who adhere to the highest environmental standards. Where possible, components of e-waste are reused or reconditioned before finally being remarketed. For example, the copper, wire, and circuit boards from CRTs can be resold, while leaded CRT glass can even be broken down and used in road construction. The result is a typical reverse supply chain process (Exhibit 8-2).

In an effort to achieve their landfill "waste diversion" targets, the City of Vancouver currently provides a small fleet of trucks (and trailers) to accomplish the "reverse logistics" portion of the CRT-related reverse supply chain. Companies like eCycle have contracted with the city to recycle CRT products, including the responsibility for unloading the city trailers. In an effort to keep their limited number of trailers moving, the city has recently instituted a clause in such contracts stipulating that the time a city trailer is at a contracted recycling facility (either waiting or being unloaded) will be charged out at a rate of $60 per hour. On the inbound side, eCycle currently operates an eight-hour day (five days a week) to match the city working hours.

Exhibit 8-1 Example of an e-waste recycling facility.

Picture Source: http://www.recyclinglives.com

Reprinted with permission of Recycling Lives.

Exhibit 8-2 Reverse supply chain process.

Until recently, eCycle had been receiving two city-owned e-waste trailers per day, but that has now increased to an average of three per day (Poisson arrival pattern). The current unloading crew of two employees is able to safely unload the trailers at a rate of four trailers per day (exponentially distributed). The weight of the CRTs requires workers to use carts/dollies to adhere to workplace safety regulations. The unloading crew employees are paid an industry average rate of $24 per hour (including benefits).

A month ago, Kevin hired Aidan Wallace, a recent business school graduate, as a business analyst to help him handle the growth of the business. Since starting, Aidan has been busy seeking new markets for reselling the components from the CRTs; however, Kevin hopes that Aidan's business degree will help in all areas of the business. With Aidan in place, Kevin was finally able during the past two weeks to take his first vacation in over a year.

Today

Returning from vacation this afternoon, Kevin dropped by on his way home from the airport to check on things and look through the pile of mail sitting on his desk. Aidan was busy working on a spreadsheet when he suddenly heard Kevin yelling:

Kevin:	"Thirty thousand dollars—are you kidding me? That was our entire profit last month - this cannot be right! I can understand the idea behind charging us while their trailers are at our facility, but $30,000 in just one month is absurd. I'm going to call the city right now and tell them they have made a massive mistake."
Aidan:	"Hold on, maybe you should ask the guys unloading the trailers if it could be true. Maybe there are lots of trailers waiting because we are not unloading them fast enough."
Kevin:	"No way—I recall my unloading guys saying they can easily unload four city trailers a day and I trust they are doing just that. If we are only having three trailers arrive per day, we should have more than enough capacity to unload the trailers. In fact, they should have some time each day when they are doing nothing!"
Aidan:	"I'm not so sure about that, and it is just after 4:30 p.m. so the city offices are already closed. Why don't you drop by the loading dock tomorrow morning and ask the guys how long the city trailers are waiting? I remember calculating queuing and waiting times while earning my business degree. I'll look it up and run some numbers this evening. We can meet tomorrow right after lunch and then decide what to do about that invoice."
Kevin:	"Sounds like a plan. I still don't believe that the $30,000 bill is correct. But, I guess with this new city policy of charging us while their trailers are at our facility, we might as well investigate ways to speed things up. Renting that second dock next door has got to be less than $30,000 a month!"
Aidan:	"You've got that right. Maybe we can just hire more people to help unload. I bet that would be cheaper than having those trailers wait."

Kevin: "Good idea. Why don't you do some of that spreadsheet 'sensitivity' analysis you are always talking about on different-sized unloading crews?"

Aidan: "I could do that. I guess that as we increase the crew size, we should be able to unload the trucks faster and faster."

Kevin: "Yes, but only up to a point. I remember helping a friend move into a new place and we eventually had so many people helping to unload that we started banging into each other and having to wait. Just because two guys can unload four trailers a day, it does not mean that six guys could unload 12 trailers a day. It would probably be more like eight trailers a day. So let's assume that each additional employee would result in one extra trailer being unloaded per day."

Aidan: "Good point. I think they call that 'diminishing returns.' I'll include that concept in my 'sensitivity' analysis."

Kevin: "Great. But don't bring me a huge spreadsheet tomorrow; I just want to see the data summary and maybe a graph or two."

Aidan: "Okay, I promise to keep it clear."

Kevin: "Be sure to run some numbers related to that $30,000 invoice from the city. I want to have some solid data to back me up when I call them tomorrow. I'm sure we owe them something for their trailers waiting, but it can't be $30,000!"

Aidan: "I'm pretty sure I can calculate the waiting time and cost using that queuing stuff. I'll have those numbers for you tomorrow morning. Go home and get some rest."

Aidan returned to his desk and made the following list of questions for which he needed to find answers.

1. How long are the city trailers waiting on average per day/week/ month? (At $60 per hour, the waiting cost can be determined.) Could the $30,000 invoice for last month possibly be accurate?

2. What are the current costs of the unloading "system" (trailers versus employees) per day/week/month?

3. How would increasing the crew size impact the total costs? What is the optimal crew size?

4. What would the new monthly bill from the city be using the optimal crew size? Sensitivity analysis on the optimal solution:
   • *Wage rate*—up to what hourly wage rate should eCycle stay with the optimal crew size?

- *Truck waiting rate*—up to what hourly trailer waiting rate should cCycle stay with the optimal crew size?
- *Arrival rate*—what is the optimal crew size if the number of trailer arrivals increases?

5. What are some other process improvement alternatives that are worth investigating further?

The Analysis

Assume you are Aidan Wallace and that you need to analyze the current situation and develop some improvement alternatives prior to meeting with Kevin tomorrow morning. Use queuing theory and a spreadsheet to determine answers to the questions just listed.

Case 9

Multi-Echelon Inventory Decisions at Jefferson Plumbing Supplies: To Store or Not to Store?

Amit Eynan[†]

It was 4:55 on Friday afternoon before the NASCAR Sprint Cup Series at Richmond International Raceway (RIR), a ¾-mile D-shaped asphalt track that is known for its four 38-foot by 24-foot high-definition LED screens used to broadcast live race action. Alex Halek, Jefferson Plumbing Supplies' Inventory Manager, was looking forward to spending the weekend at the race with two of his friends whom he had not seen since last year's Sprint Cup. He returned to his office to read a few letters that were waiting on his desk before heading home to prepare for his annual rendezvous with his friends on the track.

The first letter he opened was from Faucet Masters, a long-time supplier of a specialty faucet. This was a unique faucet that did not sell at a large volume but was steadily ordered by Jefferson's loyal customers at the rate of 100 units annually. The salespeople often commented that having the faucet available in stock was crucial to retaining some customers. The letter started with Faucet Masters expressing its gratitude for the long-term business relationship with Jefferson Plumbing. The letter also announced that lately the company had adopted a new minimum-order-size policy. Consequently, faucet number 36543 should be ordered at a minimum of 130 units. This new requirement caught Alex by a surprise as it was about three

[†] Robins School of Business, University of Richmond, Richmond, Virginia, USA; aeynan@richmond.edu

times the amount he has been ordering. Through a quick rough calculation: (3+1/3)/2, Alex figured out that tripling the order quantity will raise the management cost associated with this faucet by nearly 67%. Alex was very uncomfortable with such a cost increase. Furthermore, given the current low inventory level of this item, Alex knew that a new order would have to be placed soon, and he was determined to find a resolution even sooner.

Jefferson Plumbing sells a variety of plumbing supplies at its store in Richmond, Virginia. Because space is a premium in its city location, Jefferson tends to keep only small inventories and replenish them often. With an annual holding cost of $10 per unit for faucet 36543, Alex knew that increasing the order quantity to 130 will have a drastic cost impact. At the same time, he hoped that ordering a large quantity would reduce the number of orders. At an ordering cost rate of $100 per order, that may help a little. However, he was still concerned with his 67% increase estimate. As he walked to his car, Alex tried to shift his focus to the upcoming NASCAR Sprint Cup race and catching up with his friends.

As expected, Alex and his friends had a wonderful weekend. After the race they even had a chance to stop at the Jefferson Hotel and walk up the legendary grand staircase that is believed to have been the inspiration for the one featured in the classic movie, *Gone with the Wind*. Apparently, Margaret Mitchell, the author, stayed at the Jefferson while working on her novel portraying the interesting and dynamic relationship between Scarlett O'Hara (Vivien Leigh) and Rhett Butler (Clark Gable).

After dinner, Alex drove his friends to the airport to catch their flight back home. From the highway he was able to see the big Jefferson Plumbing sign on the top of their warehouse. Alex asked himself, "Should I place large orders of faucet 36543, keep them in our warehouse at an annual holding rate of $1 per unit, and ship small batches to the store?" However, when Alex recalled that each time the store places an order from the warehouse there is a charge of $50, he was not as sure about the advantage of this approach and decided to evaluate it the next day in the office. "After all, tomorrow is another day" (Scarlett O'Hara).

Case 10

Global Pharma: Managing Uncertainty

Sourabh Bhattacharya[†] and Surajit Ghosh Dastidar[‡]

Introduction

It was a sunny day in June, and Mr. Anil Roy had just joined one
month back as Director (Operations) of Global Pharma, a Hyder-
abad-based pharmaceutical company. He was going through a file
that Mr. Ajay Kumar, the Chief Executive Officer (CEO), had sent
earlier. He noted that in the most recent year, Global Pharma's bottom line had significantly deteriorated. This was due to high demand
variability for one of the critical active pharmaceutical ingredients
(APIs). It reminded him of a similar situation that he had dealt with in
his earlier organization. However, he was apprehensive that a similar
strategy might not work here. The scale and operational complexity of
Global Pharma was quite large compared to his earlier organization.
The downside of his decision could be disastrous for him as well as
the organization.

Company Background

Global Pharma is a $2 million company headquartered in Hyder-
abad, India. It has three segments of business: generics, APIs,
and pharmaceutical services. Its operations span across the entire

[†] Institute of Management Technology (IMT), Hyderabad, India; sourabh_iitm@
yahoo.co.in, sbhattacharya@imthyderabad.edu.in

[‡] Institute of Management Technology (IMT), Hyderabad, India; sghoshdastidar
@gmail.com

pharmaceutical supply chain that includes development and manu-facture of APIs, drug formulation, and packaging. The company has 20 manufacturing plants in India and exports to more than 40 coun-tries worldwide. The majority of its revenue comes from the U.S., European and Middle Eastern markets.

The Indian Pharmaceutical Industry

The Indian pharmaceutical market size was estimated to be $11 billion in 2011–12. The market is expected to reach $74 billion by the year 2020, with a compounded annual growth rate of 15.3%. The top ten companies constitute about one-third of the market. The domes-tic market has been growing 15% per year, while the revenue of pharmaceutical multinational companies has been growing annually by 18.7%.

The Indian Pharmaceutical Supply Chain

The pharmaceutical supply chain in India consists of three dis-tinct phases. The first phase involves production of APIs, which are the primary constituents of the drug. The second phase involves for-mulation of the drug. The third phase consists of the packaging pro-cess, which involves two types of packaging: primary and secondary. Primary packaging ensures that the drug is protected against direct external influences. This type of packaging primarily consists of blis-ters, bottles, vials, or syringes.

The distribution system is highly fragmented and complex, with more than 1,500 clearing and forwarding agents (CFAs), around 60,000 stockists (wholesalers), and approximately 550,000 retail phar-macies in the country. Most of the larger pharmaceutical companies have their own distribution centers or work with CFAs. The smaller companies distribute their products through a super stockist. A large pharmaceutical company generally has ties with about 25 CFAs in order to have a nationwide distribution network. Each CFA has an exclusive agreement with the company to maintain storage and distri-bution of its stock only. The CFAs are paid on an annual basis based on the total sales of their products. The CFAs distribute their stocks to the stockists. Stockists have the option of maintaining storage and distribution of products from multiple companies, typically five to six. Payment is made directly to the company, usually after one month. The retail pharmacies obtain stocks from a stockist or substockist and resell them to final customers (patients). (See Exhibit 10-1.)

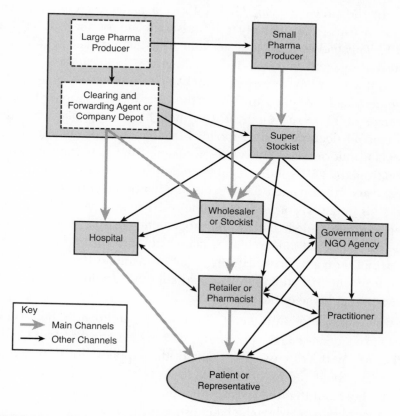

Exhibit 10-1   Patterns of distribution of pharmaceuticals in India.

The Problem

Roy was happy to be associated with Global Pharma as one of its directors. He was visibly excited by the fact that the Indian pharmaceutical industry was going through an unprecedented growth phase. However, that meant coping up with intense regulatory and inflationary pressures in addition to managing uncertainty in the supply chain.

Though proper ERP systems were in place, Mr. Roy felt that the supply chain of Global Pharma was not equipped to handle uncertainties. Such uncertainties may take the form of demand fluctuations, delayed material availability, rejection in quality, or changes in regulatory norms.

In the previous year, Global Pharma had incurred significant losses because of its inability to forecast demand for one of its critical APIs. This had resulted in a stockout situation and had significantly affected sales of the drug. The resulting cost implications were too high. Roy had to make sure that stockouts do not occur this year. The company had only one supplier for the API, and the supplier took an average of 30 days to supply it. Roy thought of postponing his order placement decision to his supplier in anticipation of more accurate demand information. He swiftly accessed some important information from the ERP solution Global Pharma had been using for the last few years. He also made a call to his supplier of the critical API to discuss his intentions of postponing the order placement decision. The supplier responded that the postponement decision would reduce the available lead time to him, which would force his production and distribution costs to escalate in order to maintain the service-level requirements. In addition, he would still take a minimum of seven days to supply the critical API.

This escalation in the cost of production and distribution would lead to an increase in the purchasing cost of the critical API for Global Pharma. As the discussion continued, the supplier also provided an estimate about the expected escalations in the price of the critical API. He stated that the escalation in the price was a function of the number of days by which he had to shorten his lead time.

Based on the information gathered from the ERP and the discussion with the supplier, Mr. Roy wondered what the optimal level of postponement should be to minimize the total annual cost of inventory management. The summary of the information collected by Mr. Roy is provided in Table 10-1.

Table 10-1 Summary of Data Collected by Mr. Roy

Variable Notation	Variable Description	Unit of Measurement	Initial Value
o	Ordering Cost	Rs. per order	100
h_0	Annual Fixed Cost of Inventory Holding	Rs./unit	1000
p	Annual Percentage of Purchasing Cost for Variable Cost of Inventory Holding	Percent	50
P	Purchasing Cost of Critical API	Rs./unit	20
$\bar{d}$	Mean Daily Demand of Critical API	Units	200
σ_d^2	Variance in the Daily Demand of Critical API	(Units)^2	100
$\overline{LT}$	Mean Lead Time of Supply	Days	30
L_p	No. of Days of Postponement	Days	0
λ	Service Level	Percent	98
s	Stockout Cost as a Percentage of Price of the Critical API	Percent	50
$\dfrac{\Delta P}{\Delta LT}$	Increase in the Purchasing price w.r.t. Unit Decrease in the Available Lead Time	Rs./day	0.15
LT_CV	Lead Time Coefficient of Variation	Percent	50

Note: Demand and lead time are assumed to be approximately normally distributed.

Case 11

Supplier Selection at Kerneos, Inc.

Ling Li[†], Erika Marsillac[‡], and Ted Kosiek[°]

Introduction

Kerneos, Inc. is the Kerneos SA subsidiary in the United States. Although headquartered in Paris, France, Kerneos SA has six manufacturing plants, one in the United States, two in the UK, two in France, and one in China. Its manufacturing facilities produce a wide range of cements based on alumina content. The lower alumina content cements, such as Fondu and Secar 41, are produced in a fusion process at two plants located in France: Dunkirk and Marseille. The higher alumina content cements, such as Secar 71 and Secar 80, are produced via a sintering process at plants in France, China, the UK, and the U.S.

Located in Chesapeake, Virginia, Kerneos, Inc. produces both high- and low-range calcium aluminate cements (see Exhibit 11-1). The high-range cements are produced in a sintering kiln located on the plant premises, while the low-range cements are produced by grinding low-range cement clinkers that are received from sister plants in France. The Chesapeake plant is capable of producing approximately 30,000 tons of high-range cements through the kiln process and 70,000 tons of low-range cements through the grinding process. Kerneos, Inc. supplies cements to customers throughout the Americas, including the U.S., Canada, Mexico, and Brazil. Customers

† Old Dominion University, Norfolk, Virginia, USA; LLi@odu.edu

‡ Old Dominion University, Norfolk, Virginia, USA; emarsill@odu.edu

° Kerneos, Inc., Chesapeake, Virginia, USA; Tkosi001@odu.edu

are supplied by either direct shipment from the plant or through a network of remote distribution centers.

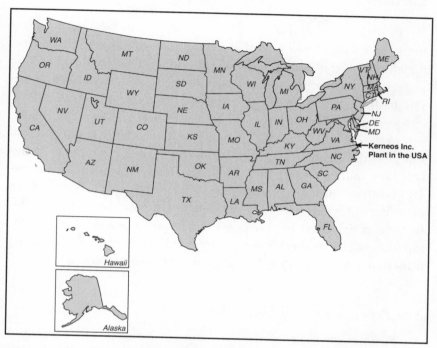

Exhibit 11-1   Location of Kerneos, Inc., USA.

In the past, Kerneos' supply chain strategy was largely based on a push approach; i.e., the plant was supplied with as much raw material as possible, and then the final products were pushed into the distribution network without consideration of demand. The plant would operate until all of the outside warehouses and onsite storage capacity was filled and would then shut down. The plant would remain closed until inventory levels dictated the need to restart. Recently, the company modified its supply chain strategy through two measures: (i) introducing new IT tools to better enable distribution and sales demand to link with production planning, and (ii) putting more emphasis on working capital.

Upstream Supply Chain for High-Range Cements

In order to manufacture the high-range cements, Kerneos has to procure various raw materials. These raw material needs include metallurgical grade alumina, and quicklime, along with other raw materials. The procurement of the alumina is centrally handled by the Purchasing department at the corporate office in Paris. This is done to leverage the alumina requirements of all of the sintering plants within the company. Contracts are negotiated with the major alumina suppliers throughout the world, and Kerneos, Inc. has input as to the suppliers' annual requirements. The alumina is shipped in a bulk vessel and the material is transported to Elizabeth River Terminals (a subsidiary of Kinder Morgan) located in Chesapeake, Virginia. The material is unloaded into bulk tank trunks, and trucks haul the material to the Kerneos site 1.5 miles away (see Exhibit 11-1). The alumina is then blown into a silo in preparation for the sintering process.

The Procurement of Quicklime

The procurement of the quicklime is handled and purchased locally by the Purchasing Manager at Kerneos, Inc. The Purchasing Manager receives input from the Production Manager regarding the quantity of quicklime needed for its annual production. The Purchasing Manager then negotiates contracts with at least two suppliers to ensure adequate supplies. The quicklime is trucked to the plant in bulk trucks and pumped off into a silo in preparation for the sintering process. Other raw materials arrive in bulk railcars, and are then pumped into silos.

The company uses a list of four major criteria to evaluate suppliers each year: quality, delivery, service, and price. The detailed subcriteria appear in Table 11-1. At the end of each year, the Purchasing Manager forms a committee of five people to evaluate the three vendors with which Kerneos, Inc. contracted in the past year. Based on the ratings, the Purchasing Manager decides which suppliers should be offered a new contract and which ones should be replaced. The Purchasing Manager has collected three ratings for each vendor, using the rating criteria listed in Table 11-1. A total of nine surveys for the past year have been collected and are presented in Table 11-2. The Purchasing Manager would like you to compute the average score, the weighted score, and the total weighted score for each vendor.

Table 11-1 Supplier Evaluation Criteria

	Criterion	Average Score	Weight	Weighted Score
	Quality		25%	
Q1	Quality problem statistics			
Q2	Quality stability			
Q3	Quality reliability			
	Delivery		25%	
Q4	Delivery statistics			
Q5	Comply with periods			
Q6	Length of term			
	Service		20%	
Q7	Enthusiasm			
Q8	Capacity of meeting emergency requests			
Q9	After sales service			
	Price		30%	
Q10	Quotation			
Q11	Competitive pricing			
Q12	Payment term			
			Total Weighted Score	

Table 11-2 Supplier Survey Data

Criterion		Vendor 1 Scores			Vendor 2 Scores			Vendor 3 Scores		
		Rater 1	Rater 2	Rater 3	Rater 1	Rater 2	Rater 3	Rater 1	Rater 2	Rater 3
Quality										
Q1	Quality problem statistics	3	4	4	4	3	5	3	4	4
Q2	Quality stability	4	5	3	3	3	4	5	5	5
Q3	Quality reliability	4	4	3	4	3	4	3	3	4
Delivery										
Q4	Delivery statistics	5	5	3	4	4	5	3	4	4
Q5	Comply with periods	4	5	4	5	5	5	4	3	3
Q6	Length of term	5	4	4	4	4	5	3	4	3
Service										
Q7	Enthusiasm with service	4	3	5	3	2	4	4	4	4
Q8	Capacity of meeting emergency	3	3	4	4	4	5	4	3	5
Q9	After sale service	4	3	5	4	3	5	4	5	4
Price										
Q10	Quotation	5	4	5	5	4	4	4	4	4
Q11	Competitive power of price	5	5	4	4	4	3	5	5	4
Q12	Payment term	5	4	5	5	4	3	4	4	4

Questions

1. Graph the supply chain structure of Kerneos SA.

2. Describe and graph the purchasing process at Kerneos, Inc.

3. Of the three vendors that Kerneos, Inc. currently uses, which vendor performed best? Which vendor should be replaced? Your recommendation should be supported by the average score, the weighted score, and the total weighted score of each vendor.

4. Are there any additional measures that you think should be considered in the decision process to keep or replace the current three vendors (other than the current scoring items)? Your suggestion(s) should be based on (a) the nature of the product, (b) the market and the supply chain structure of Kerneos, Inc., and (c) the supplier-buyer relationship management involved in supply chains.

Case 12

The Interface between Demand Management and Production Strategies at TractParts

Abhishek Shinde[†] and Dileep More[‡]

Mr. Alex is the head of operations at TractParts Pvt. Ltd., a well-known Indian manufacturer of pumps, engines, electric motors, and transformers. The company is one of India's earliest industrial groups, established in 1980, which has grown to become a big player within Indian manufacturing. Under the supervision of Mr. Alex, TractParts has been supplying modern tractor engines to leading tractor manufacturing firms such as Agri-Tractor, Ltd. and Mobile-Tractor, Ltd. The bulk of demand for tractors comes during the months from October to March. The demand for tractor engines fluctuates throughout the year. Table 12-1 shows the orders placed with TractParts by the firms Agri-Tractor, Ltd. and Mobile-Tractor, Ltd. in advance, based on their own forecasting models.

Table 12-1 Demand Pattern for Year 2003–2004

Month	Agri-Tractor, Ltd.	Mobile-Tractor, Ltd.	Total
April	500	400	900
May	300	200	500
June	100	150	250

[†] Indian Institute of Management, Calcutta, India; abhishekjs11@iimcal.ac.in

[‡] Indian Institute of Management, Calcutta, India; dileep_more@iimcal.ac.in

Month	Agri-Tractor, Ltd.	Mobile-Tractor, Ltd.	Total
July	125	100	225
August	200	150	350
September	300	350	650
October	1500	1450	2950
November	3000	3200	6200
December	3200	3500	6700
January	3800	3500	7300
February	2200	2150	4350
March	2200	2400	4600

To meet demand, the company can follow either a *chase strategy* or a *level strategy*. With a chase strategy, monthly production mirrors the total demand in that month, and to follow this strategy, the company can hire new employees during higher demand periods and lay off employees during lower demand periods. This strategy significantly reduces inventory carrying costs. On the other hand, with a level strategy, the company always produces an amount equal to the average demand per month. Excess units are stored as inventory, while all stockouts are backlogged and supplied from the following months' production. This strategy particularly saves on employee hiring and layoff costs. Moreover, to assure equal treatment, backorders are allocated to customers in direct proportion to the orders that they placed.

Currently, there are 100 workers in the TractParts manufacturing facility, and a total of 200 working hours (8 hours/day × 25 days/month) are available per worker per month. Because of strict government policies, the company cannot allow overtime.

Mr. Alex is concerned about the fluctuating demand and understands that these companies can change their demand pattern based on promotions and discounts offered to them. In order to enhance the total profit, Mr. Alex wants to optimize costs while meeting the demand pattern. He has therefore asked one of his associates to find out the associated costs and the cost structure. The associate has presented the cost structure for the engine manufacturing line as shown in Table 12-2.

Table 12-2   Cost Structure

Component	Cost	Unit
Inventory holding cost	15	$/unit/month
Stock-out cost	20	$/unit/month
Hiring cost	450	$/worker
Firing cost	750	$/worker
Labor hours required	4	Per unit
Regular labor rate	4	$/hour
Beginning inventory	750	Units
Desired ending inventory	400	Units
Selling price	280	$/unit

Mr. Alex had asked the associate to explore the possibility of offering a discount to the tractor manufacturers. Mr. Alex was informed that a maximum 10% discount could be offered per company policy, and the discount would be offered for only one month. The associate also pointed out that whenever such kinds of discounts are offered, the tractor manufacturers have a tendency to order more units in that month. However, some of that demand increase merely represents a shift from later months (early buying), as opposed to representing brand new demand.

After examining some previous data, Mr. Alex has made the following observations. For Agri-Tractor, Ltd., a 10% decrease in price for a particular month results in a 45% increase in the demand for the same month followed by a 15% decrease in demand for each of the next two months. And in the case of Mobile-Tractor, Ltd., a 10% decrease in price for a particular month results in an 80% increase in demand for the same month followed by a 25% decrease in demand for each of the next two months. However, Mr. Alex also observed that if TractPart offers a discount in the final two months of the financial year, there is no change in the demand pattern since there is a very fixed demand of tractors at the end of season.

Mr. Alex faces a unique problem, as TractParts has proven itself over the years and is considered a symbol of quality, reliability, and accountability. Being a part of this value system, Mr. Alex knows that any unfair treatment of the tractor manufacturers will not be tolerated. Mr. Alex has a meeting with the company's Senior Vice President tomorrow morning. He is concerned about which production policy should be adopted and when the 10% discount should be offered. Help Mr. Alex make the right decision by focusing on the following questions.

Questions

1. Estimate the profit made by TractParts when it follows a level strategy and no discount is offered to the tractor manufacturing firms.

2. Estimate the profit made by TractParts when it follows a chase strategy and no discount is offered to the tractor manufacturing firms.

3. Estimate the profit made by TractParts when it follows a level strategy and a 10% discount is offered to the tractor manufacturing firms in October.

4. Find out the month having the peak demand in which a 10% discount can be offered to the tractor manufacturing firms when TractParts follows a level strategy. (*Hint*: Estimate the profit made by TractParts when it follows a level strategy and a 10% discount has been offered to the tractor manufacturing firms in various months.)

5. Estimate the profit made by TractParts when it follows a chase strategy and a 10% discount is offered to the tractor manufacturing firms in October.

6. Find out the month having the peak demand in which a 10% discount can be offered to the tractor manufacturing firms when TractParts follows a chase strategy. (*Hint*: Estimate the profit made by TractParts when it follows a chase strategy and a 10% discount has been offered to the tractor manufacturing firms in various months.)

Case 13

Analyzing Distribution Network Options at Remingtin Medical Devices

Yusen Xia[†] and Walter L. Wallace[‡]

The orthopedics medical device industry is fairly large, intensely competitive, and highly innovative with worldwide sales of more than $300 billion in 2011. The United States is the largest medical devices market, with estimated sales of roughly $95 billion in 2010. The aging U.S. population represents a major catalyst for demand of medical devices. The elderly population (persons 65 years and older) is estimated to grow from roughly 40 million in 2010 to 72 million by 2030, ensuring a major boost for medical devices utilized.

This is the market that Remingtin Medical Devices, headquartered in Atlanta, Georgia, has participated in for the last thirty-plus years. Remingtin, despite the maturity of the market and its product offerings, believes that its supply chain network and inventory management systems are quite unsophisticated. For the last 30 years, Remingtin has lived with legacy systems that are not optimizing inventory levels or transportation costs for U.S. markets. In an industry where the *sale* is critical, inventory and working capital optimization take a back seat to revenue and profit margins.

Inventory is held at multiple sites across the U.S. supply chain, including Remingtin's centralized distribution center in Atlanta and nearly 100 field locations, for delivery via courier to hospitals. Sales

[†] Georgia State University, Atlanta, Georgia, USA; ysxia@gsu.edu

[‡] Georgia State University, Atlanta, Georgia, USA; wlwallace@gsu.edu

representatives keep additional inventory in the trunks of their respective cars and garages. Beyond these inventory locations, each hospital will hold a consigned inventory put in place and at the expense of Remingtin. There are two avenues for inventory planning by the hospital: scheduled surgeries where there will be seven or more days to plan and obtain inventory, and emergency (trauma) surgeries where there are 24 hours or less to have inventory in place.

Over the years, Remingtin's sales reps have developed partnership relations with the surgical doctors that prefer Remingtin's orthopedic medical devices. The switching cost of changing from one medical device provider to another is somewhat prohibitive; therefore, once Remingtin locks in with a resident surgeon, it is very difficult for a competitor to sell to that surgeon.

Because there are multiple surgeons in a hospital and each may be using a different medical device provider, the inventory levels at the hospitals grow exponentially. There are over 300 various surgical implant sizes necessary for a single knee replacement surgery and about the same number for a hip replacement surgery. Each size must be included in each surgery kit to ensure adequate inventory in case of an emergency surgery. Although not as time-sensitive, even scheduled surgeries require a complete medical device kit with each possible size of implant included to be able to perform any necessary variations of a surgery.

When an implant part is used during a surgery, it will trigger a replenishment order from the National Distribution Center (Atlanta). Each implant part is valued at approximately $2,000, and a complete kit is valued at approximately $600,000. Once an implant has been used, the kit is not available for another surgery until the used part has been replaced.

The medical device industry has faced a host of issues over the last several years, including pricing concerns, health care reform, reimbursement pressures, and increasing regulatory involvement at the state and federal levels. These issues have put downward pressure on profit margins and have caused Remingtin to take stock and reconsider the systems and procedures that it has in place to support its supply chain network, inventory management systems, and transportation methods. Remingtin believes that it has delayed the inevitable long enough. Managers now realize that something must be done to optimize their service parts logistics model and to do what it takes to stop the erosion of profit margins.

Remingtin turned to UPS Supply Chain Solutions for help. UPS offered a 10-step modeling process for Remingtin's consideration:

Step 1: Evaluate initial project scope

Step 2: Gather detailed data set

Step 3: Identify improvement opportunities

Step 4: Establish assumptions

Step 5: Baseline existing network

Step 6: Develop alternative scenarios

Step 7: Run scenarios for optimization

Step 8: Rationalize new scenarios

Step 9: Compare scenarios to each other and baseline

Step 10: Present project conclusions

UPS thought that it was critical to establish a baseline of the current network and then to develop alternative strategies to be evaluated for potential cost savings. Once this step was completed, UPS would recommend new network and transportation strategies and how they should be configured in the U.S. market. This would be followed by estimating implementation costs and issues associated with the implementation. Appendix 13-1 presents the baseline information, along with three solutions that were among several solutions offered early on by UPS. Note that the reported "Average Zone" figures are used by UPS to index Remingtin's shipping costs. The larger the Average Zone, the more expensive it is for Remingtin for shipping.

Remingtin raised the question with UPS regarding the benefits versus the cost incurred with the Southeast "milk run" delivery network originating out of Atlanta. This was costing the company just shy of $1 million. The obvious advantage was delivering the orthopedics medical device kits to their own field locations (small distribution centers) within a two-hour driving radius of Atlanta (centralized distribution center) and ensuring adequate kit coverage within that radius. The cost appeared to be extremely high, despite the advantages accrued by Remingtin.

UPS pointed out that the main benefits of milk run delivery networks are the higher utilization of company trucks and the resulting reduction of transportation costs. Other benefits include a reduction of stock, both at the supplier side and at the hospital side, a critical lead-time planning window available by using the firm's own trucks, mitigation of security for high-cost inventory, and cost savings from the integration of reusable recycling containers.

Milk-run solutions will help Remingtin perform multiple drops on a trip instead of using different trucks for different drop points. Remingtin concluded that this model would help save costs while maintaining appropriate lead times. UPS expertise allowed them to work out best routings based on data given and to scrutinize how best to optimize the trucks and thus maximize profitability for their company. Remingtin must now decide whether or not to expand the concept of milk runs with new decentralized distribution centers located within high-volume markets.

Questions

1. What are the major problems faced by Remingtin?
2. With analysis, comment on the three solutions listed previously.
3. What are the major factors (both short-term and long-term) that you would consider in helping Remingtin?
4. What other solutions can you propose to help Remingtin?

Appendix 13-1: Baseline Data and Three Solutions Offered by UPS

Baseline: Single Distribution Center (National DC: Atlanta) to Replenish All Branch Locations

Small parcel; 806,123 annual shipments; $20,875,321 annual expense

- Average Zone 6.12
- Southeast milk run delivery network – $920,100 annual expense 20,131 kits + 65,213 finished goods parcels
- 100% of the SKUs

Solution 1

Two-DC Network (East/West)—Having a Second Distribution Center in LA

Transportation Analysis Results in Parcels Shipped (% From That Suggested Location)	
Atlanta	523,980 (65%)
Los Angeles	282,143 (35%)
Average Zone Decreases to 4.65	
Outbound expense savings	($712,546)
Estimated inbound expense increase	$213,000
Net transportation savings	($512,812)
Chicago estimated warehouse expense	$3,100,000

Solution 2

Six-DC Network—Having Five Additional D.C. Locations in Chicago, LA, Philadelphia, Dallas, and Louisville

Transportation Analysis Results in Parcels Shipped (% From That Suggested Location)

Atlanta	241,837 (30%)
Chicago	185,408 (23%)
Los Angeles	96,735 (12%)
Philadelphia	120,918 (15%)
Dallas	72,551 (9%)
Louisville	88,674 (11%)
Average Zone Decreases to 3.12	
Outbound expense savings	($4,301,981)
Estimated inbound expense increase	$312,871
Net transportation savings	($3,521,654)
Five additional DC estimated warehouse expense	$7,123,000

Solution 3

Having Five More DC Locations in Chicago, LA, Philadelphia, Dallas, and Louisville, Including Milk Run Analysis

Transportation Analysis Results in Parcels Shipped (% From That Suggested Location)

Atlanta	241,837 (30%)
Chicago	185,408 (23%)
Los Angeles	96,735 (12%)
Philadelphia	120,918 (15%)
Dallas	72,551 (9%)
Louisville	88,674 (11%)
Average Zone Decreases to 3.12	
Outbound expense savings	($5,601,340)
Estimated inbound expense increase	$312,871
Net transportation savings	($4,387,432)
Five additional DC estimated warehouse expense	$7,123,000

Case 14

NunaSacha: A Facility Redesign in the Ecuadorian Andes[1]

Verónica León B.[†], Daniel Merchán D.[‡],
Ximena Córdova V.[°], Carla Tejada L.[°°],
Giuseppe Marzano[°°°]

Introduction

It was a rainy Tuesday afternoon. Gabriela Santana Flores was staring at the report she received last week. Suddenly, she was distracted by a desperate call from her young assistant, Paola. "All of our work from last week is ruined; the batch we were packing is completely contaminated! Our client will not wait any longer, what are we going to do?"

[1] Professors Verónica León, Daniel Merchán, Ximena Córdova and Research Assistant Carla Tejada of Universidad San Francisco de Quito (USFQ College of Engineering) prepared this case in collaboration with Professor Giuseppe Marzano (USFQ Business School). The authors would like to thank Paula Crespo (USFQ) and the CTT-USFQ office for their valuable assistance throughout the project. This case is based on a field project carried out by the authors and has been developed solely as a basis for academic discussion; it is not intended to illustrate effective or ineffective practices. The authors have disguised names and other identifying information to protect confidentiality.

[†] Universidad San Francisco de Quito, Quito, Ecuador; vleon@usfq.edu.ec

[‡] Universidad San Francisco de Quito, Quito, Ecuador; dmerchan@usfq.edu.ec

[°] Universidad San Francisco de Quito, Quito, Ecuador; xcordova@usfq.edu.ec

[°°] Universidad San Francisco de Quito, Quito, Ecuador; carla.tejada@estud.usfq.edu.ec

[°°°] Universidad San Francisco de Quito, Quito, Ecuador; gmarzano@usfq.edu.ec

Gabriela was head of operations at NunaSacha Export (NSE), a nonprofit organization that turned natural Andean raw materials into finished goods such as cosmetics and packaged food. Problems with contaminations and backorders had lately become too common, thought Gabriela.

The report she was reading and another call she had received the previous week made her very uneasy. An international donor group was interested in helping NSE increase its operational efficiency by investing a significant amount of money to enhance current facilities. However, two conditions had to be met: NSE had to submit a technical study to support the investment, and any enhancement had to be made within the next four months; otherwise, funds would not be granted.

Gabriela immediately asked her assistant to prepare a layout redesign proposal not necessarily limited to current equipment and space available. This kind of financial help by foreign donors was a once in lifetime opportunity and couldn't be missed, she thought.

After two weeks, Paola handed in a proposal basically stating that the current company layout was adequate and only minor changes were needed, such as new doors and bigger lockers for workers. Consequently, the money could be invested mainly on quality control equipment. Gabriela felt strongly that these arguments were not sufficient for the donor group, nor for the NSE Board of Directors, and thus decided to call in a consulting firm for help.

Organization Background

NSE, a nonprofit organization, has been working for the last 20 years with the Kañari people, an ethnic group located in the South Andean region of Ecuador.[2] This group mainly lives in rural settings and faces challenging socio-economic conditions. Therefore, any initiative aimed to support and provide assistance to this group has had a significant impact on the quality of life of this community.

---

[2] Ecuador is a country in South America with 14 million people.

In the early stages of this project, in 1984, NSE focused its activities on providing technical assistance to enhance the Kañari agricultural know-how. Since January 2002, Gabriela Santana was appointed General Manager of the organization. Beginning at that stage, NSE's goal also included building a storage facility capable of gathering all the products and then distributing them to local markets. By 2004 Gabriela had realized that by processing some of the raw materials into finished products, such as natural cosmetics and packaged food products, there were not only business but also cultural opportunities, including taking the Kañari culture overseas.

By mid-2005, Gabriela was able to get enough financial resources to install two production laboratories, one for cosmetics and one for packaged food items, and NSE launched two products: Melissa (*Melissa officinalis*) Shampoo and Toasted corn (*Zea mays*).

Some European, North American, and Asian market segments appreciated the fact that these products came from Ecuadorian indigenous communities and were in line with Fair Trade regulations, increasing their perceived value.

Over the next couple of years, Gabriela introduced two more products to the existing lines: lemon verbena (*Aloysia citrodora*) infusion and cinnamon (*Cinnamomum*) soap bar. At present, NSE has expanded its product portfolio with innovative combinations of essences for cosmetics and flavors for food products. (See Table 14-1.)

Table 14-1   Products List

Food Products		Market	
Product		Local	External
Bulk			
	Corn	70%	30%
	Coffee	10%	90%
Packaged			
	Toasted corn	32%	68%
	Infusions (tea bags)°	28%	72%
	Infusions (100gr bags)°°	42%	58%

Cosmetic Products

Product		Market	
		Local	External
Shampoo	Melissa	25%	75%
	Chamomile	11%	89%
	Verbena	65%	35%
Soap			
	Cinnamon		100%
	Chamomile	27%	73%
	Verbena	48%	52%
	Mint	31%	69%
	Lemon Verbena	33%	67%

*Cinnamon, chamomile, verbena, lemon verbena, melissa, mint, camellia (black tea), chamomile with honey, chamomile with cinnamon, verbena with mint, horchata.
**Horchata, chamomile, verbena, lemon verbena, mint, melissa.

Organizational Structure and Facility Layout

By 2009, NSE's team included 17 people among workers and specialists. The organization was divided into four functional areas, each with its own coordinator: operations, logistics, technical assistance, and administrative, plus 12 workers who were in charge of all processes (Exhibit 14-1). Additionally, NSE often hosted volunteers from all over the world who helped the organization by coordinating and providing technical assistance.

At present, NSE's infrastructure has grown into a two-floor building. There are three production laboratories, one main warehouse, a drying area, a cosmetics storage room, and three administrative offices in the basement (see Exhibits 14-2, 14-3, and 14-4).

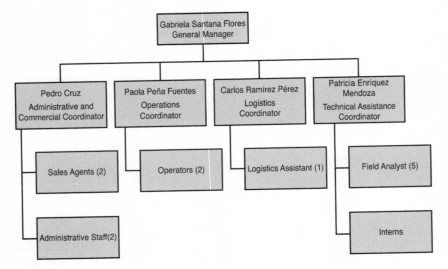

Exhibit 14-1   Organizational chart.

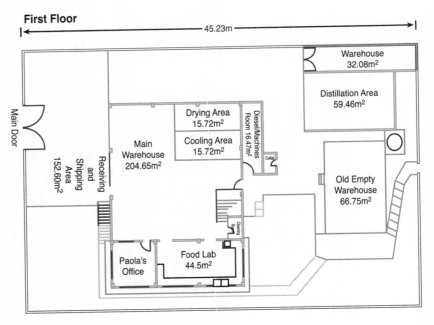

Exhibit 14-2   Existing layout, first floor.

Second Floor

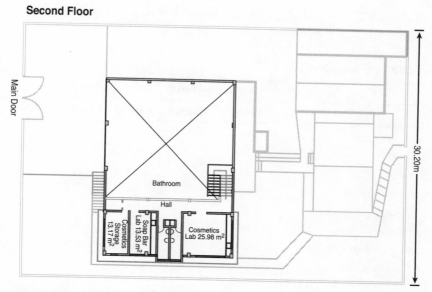

Exhibit 14-3   Existing layout, second floor.

Lateral Elevation View

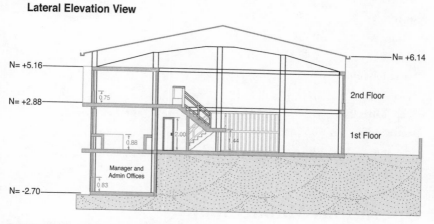

Exhibit 14-4   Existing layout, lateral view.

Even though products were successful in the market, productive conditions were not the best. The lack of space combined with the continually increasing demand caused chaos and inefficiencies. Moreover, new equipment, which had been donated since 2005 (see

Table 14-2) accounted for 20 machines that could be relocated more efficiently.

Table 14-2 Machines List

	Dimensions (in Meters)			Laboratories		
	Width	Length	Diameter	Food	Cosmetic	Soap
Shaker	1.44	1.85		x		
Oven	1.16	0.75		x		
Balance	0.25	0.32		x		
Sealer by heat	0.43	0.5		x		
Vacuum packager	0.4	1.46		x		
Grinder	0.32	0.45		x		
Soap packager	0.51	0.82				x
Turboemulsifier	1.2	0.6			x	
Stove	0.55	0.4			x	
Percolator	0.3	0.32			x	
Steel bucket	0.43	0.46			x	
Autoclave			0.45		x	
Stirrer	0.32	0.22			x	
PH-meter	0.13	0.24			x	
Sealant	0.42	0.42			x	
Mixer for liquids			0.68		x	
Small cutter	0.36	0.36				x
Soap matrix (x13)	0.77	0.43				x
Matrix sides	0.77	0.43				x
Soap mixer	1.3	1.1				x

Operations Management

By 2009, NSE had two product categories: food and cosmetics. Both of them were made with organic raw material, such as herbs, plants, and grains harvested in the Kañari Mountains by the local people (see Table 14-3):

1. **Cosmetic category:** Shampoo based on a variety of herbs and soap bars made with three mixes of herbs and roots.

2. **Food Category:** Packaged food, which includes vacuum-packed toasted corn and infusions. These products were sold in two presentations depending on the market (national or international); bulk corn and bulk coffee are only for export.

Table 14-3 Natural Herbs

	Shampoo	Soap	Infusions
Cinnamon		x	x
Chamomile	x	x	x
Verbena	x	x	x
Lemon verbena		x	x
Melissa	x		x
Mint		x	x
Camelia			x
Fragrang mauve			x
Plantain			x
Llanten			x
Amaranth			x
Escancel			x
Ataco			x
Dandelion			x
Cat's claw			x

Cosmetic Products

The production processes are described in Exhibit 14-5.

Shampoo of a Variety of Herbs

Since 2005, the shampoo production process has changed as it has adapted according to new machines, new formulas, and other operational aspects. Most of the time, two operators were exclusively devoted to this production process.

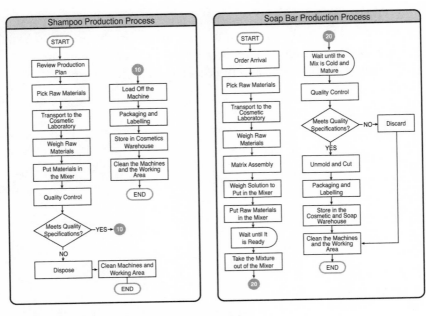

Exhibit 14-5 Cosmetics production processes.

Soap Bars

Producing soap bars proved to be more complicated than shampoo production. Nevertheless, Paola has decided that only one operator was enough for this process because this room was very small. Due to quality and presentation specifications, this area had a limited amount of light.

Food Products

Production processes are described in Exhibit 14-6.

Products Sold in Bulk

For each presentation type, the packaging for raw corn and coffee was part of the receiving process. Moreover, activities such as quality control, drying, and bagging were also performed by the operators in this process. Usually, two operators would perform these activities.

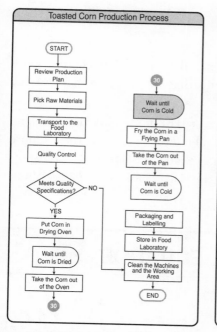
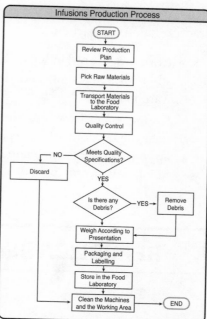

Exhibit 14-6 Packaged food products production processes.

Packaged Food

In general, the operator in charge of the soap bar production was also responsible for the packaged food products. This is why both processes were never done simultaneously.

All operators were responsible for all kinds of activities related to transformation, reception, storage, dispatching and any paperwork related to these products. In addition they were in charge of non-production related tasks such as uniform washing and machines cleaning.

Supply Chain Management

Procurement

Raw materials come from several local indigenous communities, some of them in voluntary isolation. This is a critical aspect in NSE's business mission; however, it creates significant logistical challenges due to the sometimes limited or nonexistent communications and transportation infrastructure.

- Corn is delivered in bulk. Generally, the producers are responsible for harvesting and delivering the bags directly to the warehouse.
- Coffee comes from a southern area of the Ecuadorian highland near the border with the coastal region, where weather is more appropriate for sowing and harvesting. NSE provides transportation services as part of its technical assistance (see Exhibit 14-7).

Exhibit 14-7 NSE location.

Source: Instituto Geográfico Militar, Ecuador

- Herbs, flowers, leaves, and roots come from different locations. If the community is in the proximity of the warehouse, suppliers deliver them directly. However, for faraway communities or those that are in voluntary isolation, the scheme is similar to that of coffee.

Besides raw materials, NSE needs to buy chemical supplies for cosmetics production, packaging materials, cleaning supplies and tools. All suppliers are local, and orders are delivered as required.

The warehouse does not establish time windows for receiving raw material and, as a consequence, producers and suppliers arrive at their convenience during regular operating hours.

Receiving Operations

Receiving raw materials uses a standard procedure. Basically, when a supplier arrives at NSE's warehouse to deliver raw materials, the first available person unloads and weighs the items. If materials need humidity control, the worker retrieves the appropriate instrument located in the production office; if not, only a visual quality control is performed. The accepted products are batched and taken to any storage location that is available.

Anytime raw materials arrive at the warehouse, workers have to fill out a form where they state the date, quantity, supplier region, and quality control.

Corn and Coffee

Corn and coffee have their own receiving and storage process due to quality and volume considerations. As the products arrive and batches of 200 pounds are completed, the product is temporarily bagged and stored anywhere available.

When a 200-pound batch becomes ready, two workers shell the corn and prepare the trays for drying in the oven. Coffee is directly taken to the oven. After drying, two workers weigh the products and place them into the industrial sifter. Once the product is sieved, it is then bagged and taken back to the main warehouse. These bags are specially labeled according to supplier region, date, and weight, since some international customers require corn or coffee with special characteristics regarding color, size and origin. All these activities can be performed on the same day or over several days, depending on the product arrival rate.

As cleanliness in the food and cosmetic industry is mandatory, Gabriela makes sure that everyone in the warehouse that could potentially have contact or receive materials is aware of the cleaning area procedures.

Warehousing and Picking

An average of 24,000 kilograms of raw food material is stored in the main warehouse, of which 60% is used in food products and the remaining 40% for cosmetics. By 2004, Gabriela had installed six pallet racks donated by an NGO (Non Governmental Organization) the previous year. With this equipment, NSE can store larger quantities of bulk products. They are also able to purchase chemical supplies in greater quantities and take advantage of supplier price discounts. Most importantly, finished food products can be conveniently located and organized by type, size and production date—or, at least, that is the plan.

As it turns out, despite the main warehouse being labeled per zone and per category, products in racks do not match with labels. Also, on the first floor some racks are empty, chemical products are on the floor, and cosmetics are assigned to another storage area on the second floor. Packaging materials, cleaning items and other unused equipment are also located in the main warehouse in disorderly stocking position.

Food Products

When a production order arrives, raw materials are picked up from the main warehouse and taken to the corresponding production area. Once the product is ready, the worker in charge takes the bag or carton to his or her storing area. Food products such as herbal infusions and toasted corn are stocked in the same room where they are produced (if space is available) or to the main warehouse when the batch is too big. Unprocessed products such as green coffee beans are always stored at the main warehouse and picked up from there to fulfill a pending order.

Cosmetic Products

The processes for cosmetics warehousing and pickup are a bit different, as this product category has a special storage room. A closed room on the second floor of the building is adapted to store finished cosmetics as well as small labels for exporting cosmetics and food products. Windows in this room have to be covered to ensure correct illumination conditions. Three small shelves are installed. This room is located in the end of the hallway, so the workers find circulation to and from the venue to be low.

Shipping Operations

NSE has established neither a specific shipping area nor a process. The activities are performed as customer orders arrive in any space available close to the main door.

- If the order contains only food-processed products, the worker has to go to the production room and pick the first available batch to complete the order. If there are not enough products in there, the worker has to look for it in the main warehouse.

- If the order includes a bulk product, such as coffee or corn, the product is picked up in the main warehouse. Sometimes, small bags of toasted corn are stocked in plastic containers inside the production room.

- When the order included cosmetics, the worker in charge consolidates the order directly in the cosmetics storage room and walks down the stairs with the box to dispatch it. If the box is too heavy to be carried by one single worker, then a forklift is used to take the package down.

After any of these various scenarios, a shipping order form has to be filled out and handed to the person responsible for invoicing, who is located in the administrative office.

Distribution

The distribution process varies depending on the commercial channel:

- *Local Market:* NSE currently has one store located in Cuenca, the largest city in the southern region of Ecuador located 60 kilometers south of the warehouse. A delivery truck is sent for replenishment every two weeks.

- *International Market:* This is the most important channel for NSE, representing 80% of its net revenue. Every week a truck leaves the warehouse to reach the closest international airport in Ecuador, located 250 Kilometers west in Guayaquil. An FCA[3] sales contract has been established with its international customers.

Two Different Perspectives

Paola could not stop thinking that production processes at Nuna-Sacha Exports were well-designed and that the facility did not need to be redesigned or expanded. Her belief was that higher efficiency could be achieved and cross-contamination problems would be solved if workers organized their daily activities better and if minor architectural changes were made. Nevertheless, Gabriela was still restless thinking about the international donor requirements, operational inefficiencies, and customer claims.

The consultant team received Gabriela Santana's call with enthusiasm and asked her to prepare an initial request for proposal (RFP). An extract of the RFP is presented next:

At present, our operations are functioning in an area of 1,365.95 square meters. Only 30% of the space is used for production and storage. According to some demand studies we did last year, we expect to increase the number of SKUs (stock keeping units) by 30%. In the main warehouse, we store all raw materials and food products in double reach pallets racks. The racks have four levels each, and they can handle at least eight bulk bags and 12 boxes of finished goods. If there is no space when a significant amount of raw product arrives, we place it on the

---

[3] Free Carrier (FCA) is an international commerce term published by the International Chamber of Commerce (ICC). It is a type of transaction where the seller is responsible for arranging transportation to a specific carrier at a named place.

floor temporarily. Plastic curtains keep products safe and clean by separating the production areas from the warehouse. For cosmetic products, we sterilize the bottles on the first floor and then take them to the second floor.

Operators wash herbs and plants in the sink in the food laboratory on the first floor, and they are left to dry there while other food products are being transformed or packaged. As you can see, we try to make the best with our limited resources; operators are multifunctional.

We have identified that in order to improve our quality control, we would need three new laboratories: quality control, microbiology and new products design laboratory. Please consider them in your new layout proposal.

If you have any additional suggestions, we would be happy to consider them if they are aligned to Good Manufacturing Practices (GMP) certification requirements. Furthermore, it is important to consider that to avoid production interruptions, we are not willing to stop or change our production plans.

When the consultant team read this report, they were intrigued about what else was happening at NunaSacha Export. They needed to analyze the situation from different points of view. For instance, is there a way that the production processes could be improved? Is it vital for the facility layout to be redesigned, expanded, or relocated? Considering procurement and delivery, is NSE working in the best possible way? Are the warehousing operations being handled in the most effective and efficient manner? And finally, what makes NSE unique? What about its mission provides advantages and disadvantages in the marketplace? Would any proposed solutions be different if NSE were a for-profit firm with a more traditional strategy? With all these questions in mind, the team members decided to pack their backpacks and travel directly to the site to get further information for their study, taking advantage of the trip to taste a little bit of local gastronomy—so famous in the country.

Case 15 ————————

Sherman's Supply Chain Challenge: Stopping the Retailer from Overcharging for Soda

Chuck Munson[†]

Shawn Sherman, owner of Sherman's Soda, has reached his wits' end. As producer of the region's most popular liquid beverage, Sherman's Stay-Awake Soda, Shawn has a supply and demand problem. State U. college students, comprising his largest customer base, have been complaining about his high soda prices; meanwhile, Shawn is stuck with excess capacity and excess inventory because his product isn't moving fast enough.

A couple of years ago, Shawn signed an agreement with a regional grocery chain to be the exclusive retailer for Stay-Awake Soda. At the time, Shawn was happy to only have to deal with one distribution channel, and in return the retailer agreed to provide priority shelf space and joint marketing services. The variable production cost per case of 24 cans of soda was $2.00, so Shawn decided to offer the soda to the retailer for $3.00 per case—a 50% profit margin seemed pretty good to Shawn.

After an initial start-up period involving numerous promotion activities and free giveaways, the grocery chain settled on a sales price of $6.00 per case. Sales were brisk indeed, and Shawn ramped up production to meet the stabilized demand of 3,600 cases per month. Shawn was initially happy to be earning $3,600 per month, but he was

[†] Washington State University, Pullman, Washington, USA; munson@wsu.edu

bothered that the retailer was doubling the wholesale price and earning three times the profit that Sherman's Soda was earning. It was Shawn, after all, who had created the magic addictive and tasty formula that kept his customers alert during those early morning classes and late-night study sessions.

One day, Suzy Walsh, marketing director for the grocery chain, gave Shawn a call. "Students love your drink," remarked Suzy. "But I don't think that we're making all of the money from them that we could. I'm going to keep raising prices each month for six months to see what happens. Once we locate the best price point, we'll both be better off." Shawn hesitantly agreed. He, too, was curious to know how much consumers would be willing to pay.

Table 15-1 shows the retail prices for Stay-Awake Soda in the ensuing six months, along with the result demands.

Table 15-1 Retail Pricing Test for Stay-Awake Soda

Month	Retail Price	Demand
October	$7.00	3,200
November	$7.50	3,000
December	$8.00	2,800
January	$9.00	2,400
February	$10.00	2,000
March	$12.00	1,200

At the end of March, Suzy observed the results and determined, "Demand is clearly price-sensitive for this beverage. It looks like my revenue will be maximized if I charge $7.50."

So for the next two months, monthly demand dropped by 600 units from its original stabilized point. But then Suzy's deputy director, Carol, reminded her that they should be trying to maximize profits, not just revenues. As a recent college graduate, Carol helped Suzy realize that a price point of $9.00 would be best. So the June price was set at $9.00, and the retailer's profits grew from $13,500 to $14,400 per month. Suzy was happy, and she notified Shawn that she had pinned down a final permanent retail price of $9.00.

Shawn, however, was fuming. He had sat back for nine months during this pricing test, making less money than before and watching the retailer's profits skyrocket. "This is ridiculous!" said his wife at dinner. "You should be in charge of your drink, not Suzy. Why is her company making all the money? If customers are willing to pay $9.00, then you need a bigger piece of that action. Go dust off your economics book, and figure out a proper wholesale price to fix this situation! I want you to retire in time to enjoy our dream vacation home before age 80!"

After careful study of the demand data from the past nine months, Shawn realized that his wife was right. Later the next day, he notified Suzy that he was raising the wholesale price to $8.50. "That'll show her," he mumbled to himself. Suzy was stuck. Carol explained that they could still make money by selling the soda, but they'd have to raise the retail price to $11.75 to maximize profits. "Well, now we'll be earning a stable $4,225 per month," noted Carol. "Our windfall was nice while it lasted, but the supplier caught on. It is Shawn's recipe after all."

So, Sherman's Soda was now earning a hefty monthly income of $8,450, but Shawn was still bothered. He didn't like the fact that customers were paying so much more for his drink than for the national brands, and production had dipped to 1,300 units per month. After some analysis, he decided, "I can double sales volume if the retailer will lower its price to $8.50. I'm going to lower my wholesale price to $6.75. Then I'll ask Suzy to lower the retail price to $8.50. The grocery chain will surely be willing to agree because this will raise its monthly profit to $4,550, and I'll be making $12,350 per month. Customers will be happy with the lower prices as well. This is a win-win-win for everyone!"

Unfortunately for Shawn, Suzy did *not* subsequently lower the price to $8.50 following the wholesale price decrease. Instead, she only lowered the price to $10.88. Demand rose, but only to 1,648 units. The retailer earned $6,806, but Sherman's profit fell to $7,828. Looking at the figures from July to August, Sherman's Soda did worse, the grocery chain did better, and the supply chain as a whole did better. It appeared that the only way to help the supply chain and help his customers was to hurt himself. Shawn was frustrated. He wanted to lower prices for consumers and increase overall demand, but when he tried to do that by lowering the wholesale price, the retailer gained, and the retail price didn't drop nearly low enough.

So, Shawn reluctantly raised the wholesale price back up to $8.50. But he's left with a nagging feeling that both firms are leaving money on the table. What is the best retail price and resulting monthly demand? And how can Shawn encourage the retailer to lower the retail price to that level to boost demand without eating into Sherman's own profits? What should his wholesale price be? Is the only alternative for Sherman's Soda's to sell directly to consumers and cut out "the middle man?" Shawn doesn't have the time, resources, or energy to make that switch. On the other hand, he really wants to buy that dream vacation home before he retires.

References

Jeuland, Abel P. and Steven M. Shugan. "Managing Channel Profits." *Marketing Science*, 2(3), 1983, pp. 239–272.

Mansfield, Edwin. *Principles of Microeconomics*, 4th ed. W.W. Norton and Company, New York, 1983.

Munson, Charles L., Jianli Hu, and Meir J. Rosenblatt. "Teaching the Costs of Uncoordinated Supply Chains." *Interfaces*, 33(3), May-June 2003, pp. 24–39.

Munson, Charles L., Meir J. Rosenblatt, and Zehava Rosenblatt. "The Use and Abuse of Power in Supply Chains." *Business Horizons*, 42(1), Jan.-Feb. 1999, pp. 55–65.

4
Short but Sweet

Case 16

Ethical Product Sourcing in the Starbucks Coffee Supply Chain

Dustin Smith[†]

Introduction

Fair Trade is a movement that "seeks to empower family farmers and workers around the world while enriching the lives of those struggling in poverty."[1] Fair Trade is based on the principal of paying above the market rate for goods that are environmentally friendly and made by workers in safe conditions who are paid a livable wage. Coffee is a significant focus of the Fair Trade movement, because coffee trails only oil in global trade volume.[2] Despite high global demand, market price fluctuations can create hardships for many of coffee's small producers. In the United States, the coffee market is estimated to be over $32 billion,[3] with Starbucks being the dominant coffee retailer. With its large market presence, Starbucks has been under pressure to increase the amount of Fair Trade coffee it imports. However, doing so has drastic implications for Starbucks' supply chain as Fair Trade coffee is, by design, more expensive than similar goods.

[†] Washington State University, Pullman, Washington, USA; dustin.smith@email.wsu.edu
[1] Fair Trade USA. (2010). *About fair trade usa.* Retrieved from www.fairtradeusa.org.

[2] Global Exchange. (2011). *Coffee in the global economy.* Retrieved from www.globalexchange.com.

[3] Specialty Coffee Association of America. (2012). *Coffee facts and figures.* Retrieved from www.scaa.org.

Moving forward, Starbucks must decide whether the ethical mission of Fair Trade coffee warrants the increased procurement costs.

Overview of Fair Trade

Fair Trade began in the 1940s as a small collection of European and North American organizations that focused on aiding marginalized producers by providing a market to sell basic crafts and goods.[4] These small organizations focused their efforts on importing crafts from countries such as Angola and Nicaragua. In the 1960s, "alternative trade organizations" such as Oxfam formed in Europe and started distributing imported products through a variety of small "world-shops." The movement sought to alleviate poverty among distressed populations that some considered to have resulted from growing globalization and trade imbalances.

Because sales were confined to small retail outlets and ordering through select publications such as the *Whole Earth Catalog*, sales growth was severely limited due to a lack of market presence. In order to expand distribution, retailers required a system that enabled consumers to identify a product as ethically sourced no matter where the product was sold. To solve this problem, the Fair Trade label was developed in 1988, giving products a recognizable symbol allowing fair trade goods to be readily identified as fair trade. Fair trade began to be offered in large retailers and grocery stores, spurring a growth in sales that reached an estimated $3 billion by 2007.[5] In 1997, various Fair Trade labeling groups were combined to form the Fairtrade Labeling Organization.

Despite the original focus on crafted products, the declining demand for handicrafts in 1980 prompted a shift toward agricultural goods. Initially, coffee was the major commodity offered through Fair Trade, but it has since expanded to include other products such as tea, almonds, bananas, and olive oil. Coffee was a natural target for Fair Trade groups, as it is one of the few remaining international

---

[4] FairtradeUSA. (2010). History. In Fair Trade USA. Retrieved 9/24/2012, from http://www.fairtradeusa.org/what-is-fair-trade/history.

[5] Rando, L. (2008, May 23). Worldwide fairtrade sales up 4 percent. *Confectionary News*. Retrieved from www.confectionarynews.com.

commodities that are still produced in small estates. Prior to 1973, strict quota agreements were in place among producing countries that helped stabilize prices and keep producer margins high. The collapse of regulation in 1973 resulted in the swift entry of new growers from regions such as Vietnam. The additional production of beans caused a drop in coffee prices to record lows and severely impacted the economic livelihood of producers.

In response, Fair Trade organizations partnered with producers in an effort to protect them from uncertain swings in global coffee prices. To promote economic well-being, the fair trade system uses two mechanisms. First, coffee purchasers agree to pay a minimum price for Fair Trade produced coffee. As of 2012, this minimum is set at $1.25 per pound. This creates economic stability as producers can be assured of a guaranteed price despite swings in global markets. If global prices for coffee increase above the $1.25 minimum, purchasers agree to pay a $0.10 premium. Exhibit 16-1 illustrates the price differences between Non-Fair Trade and Fair Trade coffee from 1989–2007.

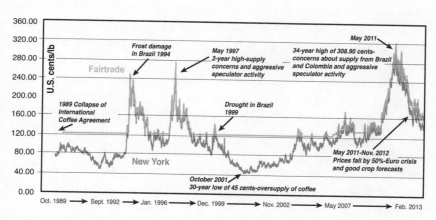

Exhibit 16-1 The Arabica coffee market 1989–2013: Comparison of Fairtrade and New York prices.

Reprinted with the permission of the Fairtrade Foundation.

Notes: NB Fairtrade Price = Fairtrade Minimum Price* of 140 cents/lb + 20 cents/lb Fairtrade Premium**

When the New York price is 140 cents or above, the Fairtrade Price = New York price + 20 cents

  * *Fairtrade Minimum Price was increased on 1 June 2008 & 1 April 2011.*

  ** *Fairtrade Premium was increased on 1 June 2007 & 1 April 2011.*

  The New York price is the daily settlement price of the 2nd position Coffee C Futures contract at ICE Futures U.S.

To be certified as a Fair Trade producer, farmers need to comply with an extensive list of production criteria, including using environmentally friendly pest control measures, enhanced storage procedures for chemicals and fertilizers, strict waste management practices, and the avoidance of genetically modified (GM) seeds for growing. For labor, Fair Trade producers are required to supply adequate personal protective equipment, prevent child labor, avoid discrimination, and provide a healthy working environment by maintaining sanitary conditions. In addition, a portion of the Fair Trade premium received by producers must be invested in the local community.[6]

The Fair Trade Supply Chain

Typical coffee production occurs on small, mostly family-owned farms, with an average size of about 2 hectares (5 acres).[7] Due to a climate favorable for coffee growing, most farming occurs in countries near the equator, with a majority of world production occurring in Brazil, Vietnam, Colombia, Indonesia, and India.[8] Coffee beans must be picked by hand; therefore, harvesting is a labor-intensive process. The ideal time to harvest coffee beans occurs when the beans have ripened into a bright red "cherry"; however, ripening times vary among plants, leading some farmers to harvest both ripe and unripe cherries at the same time. This, and other elements such as growing conditions, can create quality variations among producers. After harvesting, the coffee cherries are dried and hulled, then separated by hand to be packaged. Bulk cherries are then typically sold to a local distributor who then transports the beans to a port for export. While some producers may roast the beans prior to export, many retailers such as Starbucks and Green Mountain Coffee conduct their own roasting operations. After roasting, the prepared coffee beans are shipped to retail outlets for preparation in beverages or to be sold in

[6] Fairtrade International. (2011). *Fairtrade Standards for Small Producer Organizations*. Bonn: Fairtrade International.

[7] Calo, M. and Wise, T. (2005). Revaluing Peasant Coffee Production. *Global Development and Environment Institute*.

[8] National Geographic. (n.d.). *Major Coffee Producers*. Retrieved from www.nationalgeographic.com.

bulk directly to the consumer. Table 16-1 details estimated supply chain prices for a simplified coffee value chain.

Table 16-1 Estimated Costs per Pound for the Coffee Value Chain

Value Chain Cost per Pound	Non-Fair Trade		Fair Trade	
	Dollars	Percent	Dollars	Percent
Producer Price	$1.36	13.6	$1.56	13.6
Processing, Transport, Finance	$0.40	4	$0.40	3.4
Export Cost and Taxes	$0.10	1	$0.11	1
FOB Price	$2.59	26	$2.59	22.6
Roasting, Storage, Transport, Advertising	$5.77	58	$7.56	66
Wholesale and Retail Margin	$1.49	15	$2.29	20
Consumer Price	**$9.95**		**$11.45**	

Notes: Non-Fair Trade and Fair Trade prices estimated from the average prices of all coffee 'C' futures in 2011. Transportation prices assumed to be equal for both commodities. Wholesale and retail margins for non-Fair Trade and Fair Trade assumed to be 10% and 15% respectively.[9] Consumer price obtained from www.starbucks.com.

Of note are the higher advertising costs incurred by Fair Trade retailers vs. their conventional counterparts. The success of an eco-labeling campaign rests on the ability of an organization to educate and promote products to consumers.[10] Therefore, retailers incur additional expenses in an effort to inform the public about the virtues of Fair Trade.

Criticism of Fair Trade

Despite the good intentions of Fair Trade, some have criticized the value of Fair Trade in promoting the economic livelihood of small

---

[9] Daviron, B., & Ponte, S. (2005). *The Coffee Paradox: Global Markets, Commodity Trade and the Elusive Promise of Development*. New York: Zed Books.

[10] Rafi-USA. (n.d.). Greener fields: signposts of successful eco-labels. Retrieved from www.rafiusa.org.

farmers. The premium is pushed onto the consumer. Fair trade goods, on average, are priced 10–15% higher than other goods.[11] While consumers may be willing to pay the higher price under the assumption that the additional cost represents a charitable contribution, some evidence suggests that as little as 10% of the price premium actually reaches producers, while retailers pocket the rest.[12]

Another problem resides in consumer demand for Fair Trade coffee. In 2011, over 6.1 million metric tons of Fair Trade coffee were exported (approximately 13.5 billion pounds),[13] yet global supply far exceeded demand. For example, one Fair Trade certified producer in Guatemala reported being able to sell only 26% of its production on the Fair Trade market.[14] In these instances, farmers are forced to sell their goods to non-Fair Trade importers. Further, critics contend that Fair Trade has watered down certification standards in order to push lucrative licensing fees—fees that amounted to $6.7 million in 2010.[15] The reduction in quality, some argue, is reducing consumer confidence in Fair Trade products.[16]

Starbuck's Fair Trade Policy

With over 19,972 stores in 60 countries, Starbucks is the world's largest coffee retailer. It is estimated that in fiscal 2011, Starbucks sold close to 4 billion cups of coffee, which translates to 428 million

---

[11] Stecklow, S., & White, E. (2004, June 08). What price virtue?. *The Wall Street Journal.*

[12] Voting with your trolley. (2006, December 07). *The Economist, 381*(8507), 73-75.

[13] Fairtrade Foundation. (2011). Fairtrade foundation commodity briefing. Retrieved from Fairtrade Foundation website: www.fairtrade.org.uk.

[14] Berndt, C. E. H. (2007). Does fair trade coffee help the poor?. Fairfax, VA: Mercatus Center.

[15] Neuman, W. (2011, November 23). A question of fairness. *The New York Times*, p. B1.

[16] Haight, C. (2011). The problem with fairtrade coffee. *Stanford social innovation review*, Summer, 74-79.

pounds of coffee purchased from producers.[17] Starbucks, as part of its broader responsibility initiatives, has adopted a series of ethical sourcing principles.

According to Ann Burkhart, global responsibility manager at Starbucks, "Helping farmers thrive in the midst of a changing climate is fundamental to our mission statement and helps to secure the future of the thing we are most passionate about: incredible coffee."

In line with that mission, about 8% (approx. 34 million pounds) of Starbucks' coffee is certified as Fair Trade. In addition, Starbucks has created its own ethical sourcing guidelines called Coffee and Farmer Equity (CAFE).[18] Working with Conservation International, Starbucks has developed a series of certification standards, separate from Fair Trade, that seek to promote economic accountability, social responsibility, and environmental leadership. According to Starbucks, about 86% of all its purchased coffee complies with CAFE standards, with the goal of 100% compliance by 2015. Through the development of its own compliance criteria, Starbucks gains the ability to source from qualified but non-Fair Trade suppliers.

Calls for Greater Participation

In 2008, various consumer groups began pressuring Starbucks to increase the amount of Fair Trade certified coffee offered in its U.S. stores.[19, 20] Activists contend that since 100% of Starbucks' UK coffee is Fair Trade certified, the U.S. market is lagging. Starbucks responded by doubling their U.S. Fair Trade coffee imports to 40 million pounds in 2009 and stating that its U.S. imports are tailored to meet available demand.

Despite calls for greater participation in Fair Trade, Starbucks has pursued its own certification system, citing a lack of quality control

---

[17] Volkman, E. (2012, August 25). As coffee bean prices fall, which coffee stocks are the winners? *The Motley Fool*. Retrieved from www.fool.com.

[18] Starbucks Co. (n.d.). *Responsibly grown coffee*. Retrieved from www.starbucks.com.

[19] Organic Consumers. (n.d.). *Fairtrade vs. starbucks*. Retrieved from www.organicconsumers.org.

[20] O'Brien, C. (n.d.). *Starbucks: Fair trade or "tradewash"?*. Retrieved from www.beanactivist.com.

standards with Fair Trade coffee. However, Starbucks' corporate policies are effectively self-policing, with no third-party certification to verify its sourcing claims.[21] This leaves open the possibility of "greenwashing," whereby products are marketed as ethical yet in reality do not live up to consumer expectations of responsibility. An example of this occurred in 2006 when Starbucks began selling an exclusive Black Apron brand sourced from Ethiopia's Gemadro Estate. Starbucks claimed that the coffee was produced in accordance to the highest standards of sustainability. However, it was soon discovered that Gemadro Estate's expansion efforts contributed toward the destruction of forests and vegetation as well as threatening wildlife habitats. Additionally, Gemadro Estate was found to be part of a conglomerate that owns a variety of gold mines, oil companies, and factories.[22]

Moving forward, Ann Burkhart and the Corporate Responsibility team at Starbucks must decide how to pursue the company's ethical sourcing policies. Should Starbucks continue investing in its own standards that may face skepticism from the public? Or should Starbucks commit to sourcing more coffee from Fair Trade certified producers while running the risk that consumer demand will not keep up with supply?

[21] Jaffee, D. Weak coffee: certification and co-optation in the Fair Trade movement. 2012. *Social Problems* 59, 1. 94–116.
[22] Coffee Habitat. (2006). *Starbucks ethiopia gemadro estate: corporate greenwashing?*. Retrieved from www.coffeehabitat.com.

Case 17 ⎯⎯⎯⎯⎯⎯

Tmall, The Sky Cat: A Rocky Road Toward Bringing Buyers and Suppliers Together

Jianli Hu[†] and Olivia Congbo Mao[‡]

Company Overview

Jack Ma, a former college English teacher, founded Alibaba Group with 17 friends in his apartment in Hangzhou, China, back in 1999. His original vision was to create a 24/7 "matchmaker" B2B platform where foreign buyers could easily communicate with Chinese suppliers. Within several years the B2B platform Alibaba.com grew significantly. When eBay entered China's C2C market in 2002, Jack Ma felt the pressure immediately. He knew very well that in China's e-commerce market, the difference between businesses and individual users was quite subtle (Wang, 2010), since most individual users were also small business owners. Knowing that small and medium enterprises were the main engine of China's economy, Alibaba established the C2C platform Taobao (which means "searching for treasure" in Mandarin) to compete with eBay for the lucrative market in 2003. Despite skeptics and fierce competition, Jack Ma confidently predicted that Taobao would win the battle, "Ebay may be a shark in the ocean, but I am a crocodile in the Yangtze River. If we fight in the ocean, we lose—but if we fight in the river, we win" (Doebele, 2005).

[†] Woodbury University, Burbank, California, USA; jianli.hu@woodbury.edu

[‡] Alibaba Group, Hangzhou, Zhejiang, China; oliviam@alibaba-inc.com

With a clear understanding of the market, Taobao came with features specific to Chinese users, who at that time were still grasping the idea of e-commerce. Unlike other auction websites that charged transaction fees, Taobao's no-fee and no-threshold policy let anyone possessing a computer with Internet access to operate on its platform. Its popular "Wang Wang" instant messaging services, together with the free Alipay payment escrow system, allowed buyers to communicate with sellers before placing orders online as well as to control the authorization of payment release until they actually received the items. Users' satisfaction levels greatly improved. Although a relatively latecomer, Taobao got a quick start with its free listings and took fewer than two years to overtake eBay China, the previous leader in China's C2C market.

While occupying around 75% of the C2C market share in 2008, Alibaba saw the shortcomings of the popular auction market. Because Taobao and Alipay were both free of charge, the main revenue Taobao generated was from brand advertising and "pay for performance" in its search ranking section. Additionally, fierce competition among sellers made many merchants resort to low prices to attract customers. Taobao became a marketplace flooded with substandard products and counterfeits, and intellectual property infringement issues began to hold back the further growth of the platform.

As Chinese online shoppers' tastes and expectations grew more sophisticated, the e-commerce market in China also shifted gradually from the C2C model, where items are exchanged between individuals, to the B2C model where customers buy directly from brand owners and authorized distributors (see Exhibit 17-1). Several other e-commerce companies including 360buy.com, operated by Jingdong Century Trading Co, and Joyo Amazon, had already started their B2C operations in China. Adapting to the trend, Alibaba Group launched the B2C platform Taobao Mall in April 2008, assessed through http://mall.taobao.com, as a complement to Taobao's C2C market, Taobao Marketplace. Positioned to be a virtual shopping mall consisting of internationally known brands and major retailers, Taobao Mall aimed to raise the standards of product quality and the online shopping experience for customers. The initial fee schedules for merchants to join Taobao Mall included an annual membership fee of RMB 6,000 (approximately $940), a compulsory deposit of RMB 10,000 (approximately $1,560) that was refundable if no disputes were filed against the seller, and commissions from each transaction.

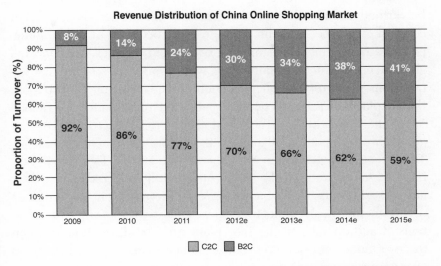

Source: iResearch Consulting Group, 2012

Exhibit 17-1 Revenue distribution of the Chinese online shopping market.

Enjoying Taobao's support, Taobao Mall started with the unrivaled advantage of access to 98 million existing Taobao consumers, and its virtual shopping mall concept quickly captured China's B2C market. Many brand name merchants tested various promotions on the platform. For example, Mercedes Benz sold 200 Smart cars in about three and half hours during its group-buying campaign on September 9, 2010. Taobao Mall's transaction volume quadrupled in 2010 from the previous year, including record sales of RMB 936 million (US $141.6 million) on its special "Singles Day" (celebrating people who are still living the single life) promotional event on November 11, 2010.

The Crisis

Compared to Taobao Marketplace, Taobao Mall possessed several advantages in attracting buyers. Merchants were required to be verified and to adhere to the standards established by Tmall that covered core aspects of the online retail process on brand building, product management, customer services, and logistics. Taobao Mall guaranteed product authenticity as well as offering a seven-day no-questions-asked return policy. To increase the traffic on Taobao Mall, the top places of Taobao's search results were reserved for the Taobao Mall sellers. In addition, Taobao Mall promised punctual delivery of goods through strategic alliances with logistics fulfillment companies. Thus, even though Taobao Mall charged fees while Taobao Marketplace remained free, small business owners still preferred to have their virtual storefronts on Taobao Mall. With the relatively low entrance requirements, many unlicensed merchants took the opportunity to register and operate on the B2C platform. The distinction between the two modes of operation soon became unclear. The difficultness of monitoring product and service quality offered by a large number of small business owners led to numerous consumer complaints, mostly about product quality and authenticity. Several foreign brands started filing lawsuits after setting up their stores on the platform. In July 2011, three Swiss watchmakers, Longines, Omega, and Rado, resorted to litigation by suing Taobao in Beijing for failing to stop the sale of knockoff watches in Taobao Marketplace, demanding Taobao to ban listings of their watches priced at under RMB 7,500 ($1,180) (Bergman, 2011).

In early October 2011, in response to the ubiquity of counterfeit and substandard products sold on Taobao Mall and in an effort to persuade small vendors to return to Taobao Marketplace, Taobao Mall issued a substantially modified new fee schedule that included an up to 10-fold increase in membership fees and 15-fold increase in cash deposits, effective in January 2012 (He & Tang, 2011). The sudden fee increases ignited a storm of online protests named "Taobao October Rising" initiated by close to 50,000 small vendors (Hille, 2011), including some who were previously fined by Taobao for selling fake products. The riots escalated when angry sellers uniformly disrupted Taobao Mall's operations through simultaneously auctioning products from major retailers, leaving negative feedback, and refusing to pay for the transactions. Several major retailers had to shut down their virtual stores temporarily. There were also thousands of sellers protesting in front of the Alibaba's headquarters building in Hangzhou, China. China's Ministry of Commerce publicly urged Taobao Mall to "listen to the opinions of all parties and take proactive measures to respond to the reasonable demands of the relevant merchants, especially small and medium-sized" (Fletcher, 2011).

Alibaba soon responded to the disputes with a revised fee schedule including a 50% reduction in cash deposit and a 9-month grace period for sellers who maintained good ratings (Table 1), as well as an investment plan of RMB 1 billion ($157 million) to make up the shortfalls in cash deposit reduction, RMB 500 million ($78 million) as a guarantee fund to help the small vendors get qualified for bank loans, and RMB 300 million ($47 million) in technical support and promotion.

The Rebranding Process

The rebranding process of Taobao Mall started well before the crisis. As Taobao's user population increased exponentially, Jack Ma pointed out that its growth would hinder the company's ability to make decisions quickly, and "we had to separate, make sure they moved faster" (Chao, 2012). Alibaba Group invested heavily on the construction of the Taobao Mall brand to better distinguish itself from the C2C portion of Taobao. In June 2011, Alibaba announced that it had decided to split Taobao into three separate units, Taobao Marketplace, Taobao Mall, and etao, an Internet search engine. Starting November 2011, Taobao Mall was renamed Tmall and began operating under an independent domain, http://www.tmall.com. Taobao explained that the move was meant to differentiate the Tmall brand and "give people a very clear idea that this is the first choice to find high-quality products" (Chao, 2010).

Table 17-1 Tmall's Fee Schedule

Item	Charge	Remarks
Monthly Fee	None	
Annual Fee	RMB 30,000 or 60,000 depends on the category	Annual fee waived if sales surpass minimum requirement in applied category and DSR rating over 4.6 within a year.
Commission	1–5% per transaction	Charges vary by product category.
Loyalty Point	0.5% per transaction	Rewards program required for all merchants.
Deposit	RMB 50,000– 150,000 depends on type of shop and trademark	Compensation for disputes occurring between merchants and customers. Full amount refunded if no disputes occur.
Product Release	None	No limit

The riots sped up the rebranding process. In January 2012, Alibaba formally changed Tmall's name in Mandarin to Tian Mao (which means "sky cat"). Two months later, Tmall came out with a new logo, a black cat drawn in the shape of the Chinese character for Sky and the English letter T (see Exhibit 17-2). The design was chosen from a contest that generated approximately 12,000 designs submitted by netizens and professional studios. The cat was inspired by the image of one appearing in ancient artwork that represented taste, wealth, and divinity. As pointed out by Tmall's President, Daniel Zhang, the ultimate goal of Tmall is "to become the 'Fifth Avenue' and 'Champs-Élysées' of e-commerce by providing the world's most fashionable, trendy, and high-quality online shopping experience." The new advertising campaign stated, "Tmall: the size of eight Fifth Avenues combined." By October 2012, Tmall has hosted online storefronts for more than 70,000 multinational and Chinese brands from over 50,000 merchants including Adidas, Gap, Mercedes, Nestle, Nike, P&G, Ray-Ban, Unilever, and Uniqlo.

Exhibit 17-2 Tmall's new logo.

Source: Tmall.com; Reprinted with permission of Alibaba

The Future

Alibaba Group's current e-commerce system includes Alibaba. com (B2B), Tmall (B2C), Taobao Marketplace (C2C), etao (product search engine), and Juhuasuan (group buying), supported by its Alipay (online payment platform), Alibaba Cloud Computing (data mining and sharing), and logistic fulfillment services (see Exhibit 17-3).

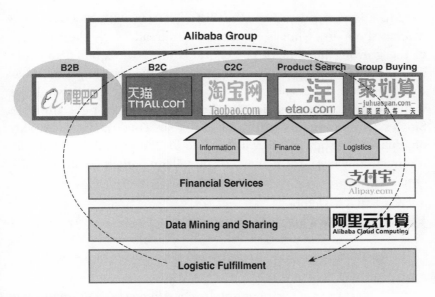

Exhibit 17-3 Alibaba Group's e-commerce system.

Reprinted with permission of Alibaba.

Although China's e-commerce market is becoming crowded with many players of various operating models, Tmall continues to be the leader in the B2C segment. It has a profit margin greater than 50%, while most other competitors only have single-digit margins (Chao, 2012). Tmall and Taobao's Singles Day promotion on November 11 has also become a phenomenon over the last several years. The one-day sales on Tmall alone climbed to RMB 3.36 billion in 2011 ($525 million) and reached to a historical high of RMB 13.2 billion ($2.06 billion) in 2012.

One of the main advantages of Tmall's operational model is that it minimizes costs and risks by not holding any products itself, while other traditional independent B2C platforms such as Jingdong's 360buy or Joyo Amazon have full responsibility of the storage and delivery of most goods. However, being just a platform also means that Tmall cannot match the consistency in sales policies and service quality provided by the traditional B2C platforms. Furthermore, Tmall faces toughening competition from traditional brick-and-mortar businesses that recently also introduced their own online shopping sites.

Most importantly, given the digitally savvy but geographically diverse shoppers, China's infrastructure already feels the pressure of the growing number of parcels flooding in the streets from the e-commerce market. Last year logistics companies complained that Tmall did not share sales information and they were overwhelmed with the sudden increase in parcel delivery requests right after the Singles Day promotion. Although this year Tmall cooperated with nine major logistics partners to prepare for the peak delivery needs, it remains unclear if the estimated 100 million parcels could be delivered on time.

Discussion Questions

1. What are the main strengths and weaknesses of Tmall's business model?

2. What strategies and actions taken by Alibaba have upset its relationship with small sellers? Were those strategies and actions necessary? Are there any better alternatives Alibaba could consider?

3. Should Tmall continue its Singles Day promotion in the future? What are the main disadvantages of the temporary discounts and their impacts on inventory management?

References

Bergman, Justin (2011). "The eBay of the East: Inside Taobao, China's online marketplace." *Time*, November 2, 2011. Retrieved from www.time.com/time/world/article/0,8599,2098451,00.html.

Chao, Loretta (2010). "Alibaba gives Taobao Mall retail site more prominence." *Wall Street Journal (Online)*, November 1, 2012. New York, NY.

Doebele, Justin (2005). "Standing up to a giant." *Forbes*, April 25, 2005. Retrieved from http://www.forbes.com/global/2005/0425/030. html/.

Fletcher, Owen (2011). "The new online battleground; For China's e-commerce sector, the focus shifts to business-to-consumer sites." *Wall Street Journal (Online)*, October 27, 2011. New York, NY.

Gao, Kane (2012). "Tmall's new logo, and its amusing heritage." *Illuminant: Public Relations and Strategic Communications*. Retrieved from www.illuminantpartners.com/2012/04/09/tmall-logo-amusing-heritage/.

He, Wei and Zhihao Tang (2011). "Taobao Revises Fee Plan." *China Daily*, October 18, 2011. Retrieved from http://www.chinadaily.com. cn/bizchina/2011-10/18/content_13921477.htm.

Hille, Kathrin (2011). "Alibaba hit by Internet protest." *Financial Times*, Oct 13, 2011. Retrieved from http://www.ft.com/cms/s/0/a849c9f8-f59b-11e0-94b1-00144feab49a.html#axzz2L2VcLY2k.

iResearch Consulting Group, China (2012), "Chart 2-1 Revenue of 2008-2015 China Online Shopping Market," *2011-2012 China Online Shopping Research*, May 24, 2012.

Wang, Helen H. (2010). "How eBay failed in China." *Forbes*, September 12, 2010. Retrieved from http://www.forbes.com/sites/china/2010/09/12/how-ebay-failed-in-china/.

Case 18

Make to Demand with 3-D Printing: The Next Big Thing in Inventory Management?

Tom McNamara[†] **and Erika Marsillac**[‡]

Imagine that you are the manager of a high-volume continuous process factory. Suddenly, your production line goes down because a machine fails. A technician tells you that the problem is a critical small part that needs to be replaced. You used to keep plenty of these parts on hand for just such an emergency. But since your company instituted a "lean" inventory policy in order to reduce the amount of material you keep on hand, you no longer have any spares. Ordering a replacement part will take several weeks, and the cost of lost production due to the line being down will run into the millions of dollars. Wouldn't it be great if there was a way to "magically" produce the part you needed? Well, soon there might be.

A technique called 3-D printing (otherwise known as additive manufacturing) could provide just such a solution. It is a fairly simple concept. First, design specifications for the item you need are fed into a 3-D printer (which often looks and works just like regular document printer). Next, the work piece gets built as layers are carefully and methodically "printed" onto a flat surface, repeatedly, until a fully formed three dimensional item appears, like adding layers to a cake. The layer materials can be as diverse as powdered metal, liquid

[†] ESC-Rennes, Rennes, France; tom.mcnamara@esc-rennes.fr
[‡] Old Dominion University, Norfolk, Virginia, USA; emarsill@odu.edu

plastic, or even specific metal-based inks, which are used to print and build electronic circuits.

The concept of 3-D printing is not new. Since the late 1980s, it has been used to manufacture individual samples or prototypes, mostly for demonstration purposes. But thanks to improved technology, lower costs, faster printing speeds and better quality, 3-D printing is finding its way into a whole host of new applications and markets that were formerly out of reach. In many cases, 3-D printers can now make things, with the same level of quality (in some cases even better), with less material, more quickly, and with fewer constraints than traditional production methods.

Traditional production of metal or plastic commercial items uses industrial dies, which stamp out what you need from a piece of metal, or injection molds, which are filled with liquid plastic to form a shape as they dry. Another traditional way to produce a commercial item is to take a block of metal and cut away at it on a lathe (otherwise known as "machining") until it is in the shape you need. These well-proven methods can be laborious, costly, time consuming, and wasteful (for example, machining usually wastes 90% of the material being used), and quite often can produce only a limited product range. They end up requiring long, high-volume production runs to make them cost effective.

Due to the lengthy lead times that can be required to manufacture an item in these traditional ways, companies normally need to keep vast inventories of critical spare parts on hand, tying up money that could have otherwise been put to better use. The advent of 3-D printing could allow companies to put an end to this practice, drastically lowering the cost of inventory by letting manufacturers produce exactly what they need, when they need it.

The main drawback of 3-D printing, however, is still the relatively high cost of the specialized printer. Industrial 3-D printers can cost well over a $1 million, although prices are dropping fast. A top-of-the line traditional milling machine, on the other hand, can be obtained for around $50,000 (with second-hand machines in excellent condition costing a fraction of that). Also, the specialized metal and plastic materials used in 3-D printing are generally far more expensive than conventional materials. In addition, there really are no cost benefits achieved from large production runs, so 3-D printing is still best suited to low volume/high customization applications.

Despite its comparatively high cost and limited applications, 3-D printing is a growing industry, with about 28% of current production resulting in final, finished products versus a history of producing only prototypes. Sales of 3-D printed products were $1.7 billion in 2011, double what they were in 2007, and they are expected to reach $5.2 billion by 2020, according to Wohlers Associates, a consultancy. This trend should only continue as printers get faster and cheaper.

Questions

1. What are some potential drawbacks and benefits of 3-D (or additive) printing?
2. Do you believe that 3-D printing has a future? Explain.

Activity

Visit the website for 3-D Systems (production3dprinters.com). What are some of the industries benefitting from 3-D printing? What are some of the types of products being made?

Sources

"3-D printing could remake U.S. manufacturing" By Paul David-son, *USA Today*, July 10, 2012. Accessed at: http://www.usa-today.com/money/industries/manufacturing/story/2012-07-10/digital-manufacturing/56135298/1.

"3D Printing Industry Will Reach $3.1 Billion Worldwide by 2016" By T.J. McCue, Forbes.com, March 27, 2012. Accessed at http://www.forbes.com/sites/tjmccue/2012/03/27/3d-printing-industry-will-reach-3-1-billion-worldwide-by-2016/.

"Additive manufacturing: Solid print" *The Economist*, April 21, 2012. Accessed at http://www.economist.com/node/21552892.

"Print me a phone" *The Economist*, July 28, 2012 Accessed at http://www.economist.com/node/21559593.

"3D Printers: Make Whatever You Want" *Bloomberg Business-Week*, April 26, 2012. Accessed at http://www.businessweek.com/articles/2012-04-26/3d-printers-make-whatever-you-want.

Case 19

Airbus' Overstretched Supply Chain: Just How Far Can You Go Before Your Supply Chain Snaps?

Erika Marsillac[†] **and Tom McNamara**[‡]

For decades, companies have been doing their best to increase the performance of, and decrease the costs from, their supply chains. As a result, more and more manufacturers are going farther afield to find suppliers who can provide them with necessary parts and materials at the best cost. Global expansion has given rise to intricate networks and complicated supplier relationships that require highly adept managers to coordinate. Because these supply chains are extremely "lean," there is very little supplier redundancy or margin for error. But when your supply chain is that lean, the cost of getting something wrong or misjudging a delivery date can be catastrophic. So this begs the question, "Just what is the breaking point of a lean supply chain?" Airbus is about to find out as it fast approaches the limits of its supply chain.

The airline industry is, by its very nature, one of boom and bust, but it seems to currently be in a growth phase. One estimate sets the global market value for aircraft (of all types) at $4.5 trillion by 2031, with many aging jets that are currently in service needing replacement. Airbus sees most of this growth coming from emerging markets such as China and India, where the rising middle class is keen to travel. In anticipation of this market surge, as well as to meet present

[†] Old Dominion University, Norfolk, Virginia, USA; emarsill@odu.edu

[‡] ESC-Rennes, Rennes, France; tom.mcnamara@esc-rennes.fr

demand forecasts, Airbus is increasing its manufacturing capacity for single-aisle aircraft at its assembly plant in Alabama in order to augment existing capacity at two other plants in France and Germany.

Airbus is facing a problem that many companies would love to have: an order backlog of 4,341 passenger jets. To meet this outstanding demand, the company believes that it must increase production. But coordinating the work of Airbus' estimated 1,500 suppliers is a Herculean task. And getting more out of its already overstretched supply chain looks to be just as challenging.

The weakest link in Airbus' operation appears to be in the production of one of its best-selling planes, the single-aisle A320. The A320 planes are presently being made at a rate of 40 per month, with output expected to rise to 42 per month by the end of 2012. Ideally the company would like to produce 44 planes per month, but there is a problem. Its supply chain just is not up to the task. A related concern is that these supply chain limitations might also hamper technical upgrades to and future versions of existing models.

Compounding this problem is the trend of supplier consolidation. Quite often, the same supplier company provides parts to multiple aircraft manufacturers. If demand suddenly increases, it can be quite difficult to quickly ramp up production. One of the negative side effects of Airbus' formerly relentless drive toward lean was that excess and redundant capacity has largely been removed from its supply chain.

This constraint doesn't reside so much with the larger tier-1 suppliers, but rather further up the supply chain with the tier-2 and tier-3 suppliers. When it comes to quick changes in output and expanding capacity rapidly, "many small companies aren't up for it," says Robin Southwell, CEO of EADS UK. One reason for this is that some key suppliers were unable to make critical and much needed investments due to the financial crisis of 2008. Their ability to react to changes in the marketplace is now suffering as a result.

Airbus, and its nemesis, Boeing, have been encouraging their larger tier-1 suppliers to purchase smaller partners in their respective supply chains. The logic behind this strategy is that the larger companies, by gaining size and scale, would be more able to adapt themselves to changes in the marketplace and be more reactive, making the overall supply chain more robust.

Airbus is no stranger to problems with its sourcing network. The company, if needed, is willing to provide substantial financing and technical support to remedy problems in its supply chain, including actually purchasing suppliers, as it has done in the past. But Airbus prefers to have suppliers consolidate among themselves. Airbus is presently hoping that the suppliers of the components that go into fuselages and wings (known as aerostructures) will come together in some kind of consolidation action. The aerostructure suppliers are in the airplane market sector that generally has the lowest profitability, so the logic exists for these companies to come together to develop economies of scale and higher efficiencies.

Tight finances and a difficult economic climate should only continue the trend of supplier consolidations and mergers, thus making Airbus' supply chain even more fragile and susceptible to the vagaries of the market.

When it comes to lean, it could very well be a situation of, "Be careful what you ask for!"

Questions

1. What are some current trends among suppliers in the aviation industry?

2. What steps is Airbus willing to take in order to rectify problems in its supply chain and help suppliers?

Sources

"Airbus and Boeing push supply mergers" By Andrew Parker and James Shotter, July 8, 2012, *Financial Times*. Accessed at: http://www.ft.com/cms/s/0/2b66574a-c73b-11e1-849e-00144feabdc0.html#ixzz25oI9XGrp.

"Airbus chief looks beyond the A380" By Andrew Parker, June 24, 2012, *Financial Times*. Accessed at: http://www.ft.com/cms/s/0/3ef56200-bde6-11e1-9abf-00144feabdc0.html#ixzz25oN508Ay.

"Airbus Developing 'Stretch Marks' as Supplier Strain Hurts Sales" By Robert Wall and Andrea Rothman, September 4, 2012, *Bloomberg News*. Accessed at: http://www.bloomberg.com/news/2012-09-04/airbus-developing-stretch-marks-as-supplier-strain-hurts-sales.html.

"Suppliers face battle to meet plane order backlog" By Victoria Bryan and Karen Jacobs, July 11, 2012, *Reuters*. Accessed at: http://in.reuters.com/article/2012/07/12/airshow-suppliers-idINL6E8IC2EP20120712.

Case 20 ——————

How to Keep Your
Food Supply Chain Fresh

Tom McNamara[†] **and Erika Marsillac**[‡]

As consumers demand greater food variety, retailers are under increasing pressure to provide their customers with a greater choice of fresh produce, including fruits, vegetables, fish, and fresh flowers. Whether it's fair trade coffee from Bali, fresh cut roses from Kenya, or Spanish cherries in winter, people expect to readily find the products they want, when they want them, and at a price they can afford.

This puts extreme pressure on supply chains as they become even more overstretched and far flung in trying to provide exotic goods. Managing this complexity is a job for experts. As information technology makes using real-time data more feasible in the decision-making process, and as forecasting and modeling become more accurate, one would expect the job of a supply chain manager to get easier. But this is far from the case, and the cost of not getting things right can be catastrophic.

It has been estimated that the losses from unsold fruits and vegetables are as high as $15 billion per year. The National Resources Defense Council (NRDC) argues that almost half of the U.S. supply of all fruits and vegetables goes to waste and is unconsumed. This is hugely inefficient and a source of much concern, especially for companies that are trying to be "lean and green." Even more troubling for the food industry, it appears that increasing government regulations

[†] ESC-Rennes, Rennes, France; tom.mcnamara@esc-rennes.fr
[‡] Old Dominion University, Norfolk, Virginia, USA; emarsill@odu.edu

are on the way. The European Union recently passed a resolution calling for food waste to be cut in half by the year 2020. Over 50 food retailers in Great Britain have already instituted their own resolutions for reducing food waste in their operations and supply chains. Where Europe goes, will the rest of the world follow?

Supply chains, for the most part, utilize one of two modes of operation. They can be considered to be operating as a *push* system or as a *pull* system. In a push system, a supplier (or producer) sends a fixed batch of fresh produce items to some wholesaler or intermediate market. A distributor then buys the items that it needs, or expects to sell, and makes them available to the general public, leaving what it does not need for other buyers. In a pull system, however, almost all of the action starts downstream with the distributor, who places an order for a particular list and quantity of fresh items. The producer then ships only the ordered items. Due to the nature of the goods, a certain amount of rot, damage, or decay is usually expected to occur. That means that when the distributor receives its shipment at the final destination, the quantity of goods available for sale could be less than was originally expected (since the supplier does not really have an incentive to overfill the order).

So, of the two available systems (push vs. pull), which one delivers "fresher" results in an operating environment that involves fragile, delicate, or perishable goods? Recent research has shown that in these types of perishable supply chains, better performance is usually achieved by using a pull type of operation. But it's a pull system with a twist—there is a special compensation mechanism for the producer.

Because of the fragile nature of the products being shipped, the producer usually carries a greater amount of risk than the distributer does. The producer usually has to ship more product than was originally ordered or paid for to ensure that a sufficient quantity arrives at the final destination. This excess can bring about inefficiencies and misaligned incentives, resulting in lower overall performance (and profits) for the supply chain as a whole.

One way to overcome this problem is by developing a negotiated compensation mechanism for the supplier. When an order is placed, the amount shipped is increased by a predetermined multiplier. For any additional material that arrives (i.e., if there was less waste, damage or spoilage in transit than expected), the supplier receives compensation for it at a discounted wholesale price, which has been pre-negotiated.

To get the best performance and highest profitability out of a supply chain dealing in perishable goods, a great deal of trust and respect is required by the respective partners. By determining a fair and equitable price for the compensation mechanism, the distributor can reduce the amount of risk for the producers and compensate them for additional inventory that would normally have gone to waste. And that means less food going to the dumpster!

Questions

1. What are some of the challenges to delivering perishable goods to customers?
2. What is the difference between the push and pull systems? Is one better than the other for this type of product?

Activity

Visit NXP's website and read their whitepaper on the use of technology to better manage the perishable food supply chain: http://www.nxp-rfid.com/sites/nxprfid.com/files/Carbon%20footprint%20White%20Paper.pdf.

What kind of technology is being used to track food and reduce waste? What does the technology do?

Sources

"America Trashes 40 Percent of its Food: Tips to Cut Waste," *Living Green Magazine*, August 23, 2012. Accessed at http://livinggreenmag.com/2012/08/23/people-solutions/america-thrashes-40-percent-of-food-supply-tips-to-cut-waste/.

"Food Waste Causes Losses Throughout the Supply Chain," August 23, 2012, *Environmental Leader*. Accessed at http://www.environmentalleader.com/2012/08/23/food-waste-causes-losses-throughout-the-supply-chain/.

"Supply Chain Management of Fresh Products with Producer Transportation," by Yongbo Xiao and Jian Chen, *Decision Sciences*, Volume 43, Number 5, October 2012.

"Wasted: How America Is Losing Up to 40 Percent of Its Food from Farm to Fork to Landfill," by Dana Gunders, The Natural Resources Defense Council (NRDC) Issue Paper, August 2012. Accessed at http://www.nrdc.org/food/files/wasted-food-IP.pdf.

Case 21

The End of Lean?: Automobile Manufacturers Are Rethinking Some Supply Chain Basics

Erika Marsillac[†] and Tom McNamara[‡]

Minimum inventories, consolidated suppliers, and just-in-time deliveries—these basic tenets have been the mantra of lean supply chain management for decades. But recent events are causing many automobile manufacturers to rethink the way they design and manage their supply chains.

For years, manufacturers have been trying to make their operations as lean as possible, stretching their supply chains almost to their breaking point. But two recent events showed managers that there just might be such a thing as "too lean." One was the devastating earthquake and subsequent tsunami that took place in Japan in March of 2011. This one-two punch natural disaster caused many Japanese automotive suppliers to go offline, sometimes for months, resulting in worldwide shortages for some key components. The other event was a devastating explosion in early 2012 at a key German chemical plant that produced a special type of resin used in fuel lines. Because many automotive companies had been studiously eliminating redundant suppliers from their supply chains to reduce complexity and costs, suddenly there was no backup plan. Tragedies such as these could negatively impact the operations of several car companies, resulting

[†] Old Dominion University, Norfolk, Virginia, USA; emarsill@odu.edu
[‡] ESC-Rennes, Rennes, France; tom.mcnamara@esc-rennes.fr

in model shortages and idle assembly lines if alternative suppliers cannot be found.

Another aggravating factor has been the downsizing and consolidation of several car component suppliers. Many of the smaller suppliers simply do not have the money, labor, and capacity to deal with last-minute orders or large fluctuations in demand. Other suppliers have gone out of business because of an inability to deal with the fast-changing competitive global landscape. To take just one example, according to the Original Equipment Suppliers Association, the U.S. has seen 57 manufacturers either close, get taken over by another company, or file for bankruptcy since 2008.

But the automobile industry is not alone in dealing with supply chains that may have been cut too lean. In November 2011, severe flooding in Thailand wreaked havoc with the supply chains of many high-tech companies. Although not all companies were directly affected by flooding in their production facilities, most found that the suppliers for their key components were. For example, Seagate, a provider of hard drives for PCs and servers, expects that disruptions to its operations might not get resolved until 2013. Apple, Hewlett-Packard, and Intel are just some of the other companies expecting negative earning impacts from the natural disaster.

Realizing the fragility of their lean supply chains and supply networks, some carmakers are thinking of implementing the unthinkable; that is, going back to the bad old days of stockpiling inventory and keeping large numbers of vital components on hand. This would overturn almost 30 years of industry practice and conventional wisdom.

With more extreme weather expected in the coming years, companies might need to rethink the balance between cost effectiveness and the potential for lost profits due to disruptions of their supply chains. But what actions are available to managers who want to increase the robustness of their facilities and their supply chains?

A possible course of action carmakers could consider would be judicious and well-thought-out increasing of inventory levels of some (but not all) critical components. They also might want to develop relationships with alternative suppliers or have several backup suppliers in place. Yes, this could increase the cost of inputs somewhat, but the possible benefits of increased customer satisfaction and reduction of lost sales might well compensate for any increase in cost. Ultimately, some manufacturers might want to bring some fabrication work back in-house, or at least develop the capacity to do some in-house work. Of course, these suggestions might well be considered blasphemy in the world of supply chain management and lean operations, but some automakers are starting to feel desperate.

Another possible solution might be to avoid the common industry practice of "clustering," whereby a group of similar industries and manufacturers locate their operations in a common geographical area. Clustering has its benefits, namely increased technological expertise, rapid learning effects, economies of scale, and economies of scope, all of which are important in today's dynamic global environment. But, if a massive earthquake or storm hits a geographically centralized cluster, a large chunk of a company's supply chain can be wiped out. Richard Little, Director of the Keston Institute for Public Finance and Infrastructure Policy at the University of Southern California, believes that "Companies need to rethink the clustering model. Yes, you get benefits but you also have common vulnerabilities for an entire industry or sector."

As always, managers will need to weigh the pros and cons of having a lean and efficient supply chain against the consequences and costs of building in some supply chain redundancies and flexibility. In the end, in order to avoid shutdowns and disruptions, those costs might be worth incurring in order to provide higher levels of customer service.

Questions

1. Briefly explain the difference between a lean supply chain and a traditional supply chain.

2. What are some of the benefits and consequences of lean operations?

3. What is "clustering"? What are some pros and cons of clustering?

Sources

"Crises make automakers rethink lean parts supplies" By Dee–Ann Durbin and Tom Krishner, April 20, 2012, Associated Press. Accessed at http://cnsnews.com/news/article/crises-make-automakers-rethink-lean-parts-supplies.

"Supply chain disruption: sunken ambitions" By Ben Bland and Robin Kwong, November 3, 2011, *Financial Times*. Accessed at http://www.ft.com/cms/s/0/6b20d192-0613-11e1-ad0e-00144feabdc0.html#ixzz25Lni2GRc.

"The global supply chain: So very fragile" By Bill Powell, December 12, 2011, *Fortune*. Accessed at http://tech.fortune.cnn.com/2011/12/12/supply-chain-distasters-disruptions/.

5
Unique Challenges from Around the Globe

Case 22

A Brazilian Dairy Cooperative: Transaction Cost Approach in a Supply Chain

Pacheco Dohms[†] **and Sergio Luiz Lessa de Gusmão**[‡]

Supply chain studies are important because a good management system and governance between its members can generate systematic earnings and competitive advantages for an entire supply chain. Therefore, understanding the way that a supply chain is structured and determining the main relations and transactions between its members are essential to making a critical analysis for systemic optimizations.

A cooperative supply chain has some extra peculiarity, based on its values of self-help, self-responsibility, democracy and equality, equity, and solidarity. In the Cooperativa Agropecuária Petrópolis Ltda. (COAPEL), a cooperative company (co-op) that operates in the Brazilian dairy industry, the managers have some doubts regarding the actions that need to be taken to achieve a better transaction efficiency to support the main activities of COAPEL's supply chain and also to minimize risks within the supply chain.

This approach in this case includes a mapping of the co-op's supply chain, the identification of its primary members, and a structural analysis of the chain. The findings presented here are based on

[†] Pontifical Catholic University of Rio Grande do Sul, Porto Alegre, Brazil; dohms.fernanda@gmail.com

[‡] Pontifical Catholic University of Rio Grande do Sul, Porto Alegre, Brazil; slgusmao@pucrs.br

transaction costs approach attributes and ways in which the management and the cooperative business model can influence them.

Background for the Cooperative and the Industry

COAPEL, commercially known as Piá (trademark of products), was founded in October 29, 1967, in the district of Vila Olinda in Nova Petrópolis in the state of Rio Grande do Sul, in the South region of Brazil. The co-op's history started with a partnership between the German government and a group of 213 farmers in the region, with the goal of developing milk production in that area and seeking more price competitiveness for their dairy products. Another target of this project was to transform farmers into agricultural businessmen through professionalization, increasing their competitive edge with a better production environment and more efficient processes.

The co-op has grown, expanding its product mix and geographic reach. It currently operates in more than 80 cities in Rio Grande do Sul state, supported by the performance of its 15,000 members and revenues of 414 million Reais. The co-op represents 8.5% of dairy sales in the South Region of Brazil (Latin Panel data). The main focus of COAPEL is the Rio Grande do Sul market, representing 65%. In addition, Santa Catarina represents 20% and Paraná 15%. COAPEL usually buys milk from its members, but sometimes it will buy milk in the "spot market" when demand is higher than the supply that its members can provide. The supplier network includes both large and small farmers from the region.

Brazil is the fifth largest milk producer in the world, behind the USA, India, China, and Russia. The recent production levels represent a huge cycle of growth based on the introduction of UHT in the 1990s, changing the way of production and also manners of consumption. Investments flowed into this area, introducing technology and modernizing production with new techniques, machinery, and equipment. Milk consumption in Brazil remains low (15 kilos compared to 37 in Holland) but with better prospects due to the growth of population income and changes in diet.

The International Cooperative Alliance (ICA) provides the following definition: "A co-operative is an autonomous association of persons united voluntarily to meet their common economic, social, and cultural needs and aspirations through a jointly owned and democratically controlled enterprise." A cooperative is also defined as a voluntary association of people, joining their production force, expenditures and, savings capacity to develop an economic and social business, generating to all members systematic gains (OCERGS, 2010). A co-op in this business model has its own values and principles. ICA has established seven principles for a co-op: voluntary and open membership; democratic member control; members' economic participation; autonomy and independence; education, training and information; cooperation among different co-ops; and concern for community.

Cooperative relationships are based on co-op's principles, giving emphasis to a long-term philosophy where everybody wins (Chistopherson and Coath, 2002). As a final point, some cooperatives' doctrines, such as democracy, are useful for understanding the relationships between its members. For example, this relates to the manner in which a cooperative should be structured (administrative and management); the degree of free membership as well as free leaving; and the return of surplus capital, after deductions.

Transaction Costs Approach

The transaction costs approach, first studied by Ronald Coase in 1937 and further developed by Oliver Williamson in 1985, is the study of economic organizations, focusing on transactions and economic efforts to accomplish their activities. These transactions, according to Williamson (1985), occur when goods or services are transferred between different interfaces, having inherent process characteristics that determine the way that outputs are generated and how they are delivered to a customer. Costs are incurred for involved members to effect transactions, and internal costs accrue for each member. These relations and transactions are influenced by several factors, categorized by Williamson (1985) as *behavioral* and *dimensional* assumptions.

The behavioral assumptions focus on understanding the way that human nature works and also how the institutions work, such as laws and society, that shape these behaviors. Transaction costs related to behavioral factors include bounded rationality and opportunism of agents. Bounded rationality is related to cognitive limits of competence to formulate and solve complex problems where the knowledge of all variables is limited in a decision process, making it difficult for an agent to choose between different alternatives presented to him. Organizations try to avoid it with governance structures predicting and anticipating transactions in-house. It is also suggested as a way to increase rationality interactions between different agents as guidelines based on groups that can generate more efficient results than individual actions can (Simon, 1971, *apud* Gusmão, 2004). Opportunism of agents is related to the pursuit of self-interest to the detriment of others seeking their own benefits.

The dimensional assumptions are related to the way in which transactions are realized, with peculiarities of each organization. The main dimensions that describe a transaction are (1) asset specificity, which describes characteristics of an asset that express its specific value and usefulness; (2) uncertainty, which describes the future risks of a transaction related to its flows, difficult to be predicted and covered by contracts; and (3) frequency, which describes a transaction recurring between two agents.

COAPEL Supply Chain

COAPEL management hired a company to analyze the supply chain, and one of the first findings was its structure, as shown in Exhibit 22-1.

After a restructuring process at COAPEL over the past few years, its activities were subdivided into two units, main unit and support unit, to facilitate a better development of activities. The main unit is comprised of the dairy industry, input factories and distribution centers, agriculture retail, and consumer retail/supermarkets. The support unit includes technical assistance, controllership, marketing, human resources, etc.

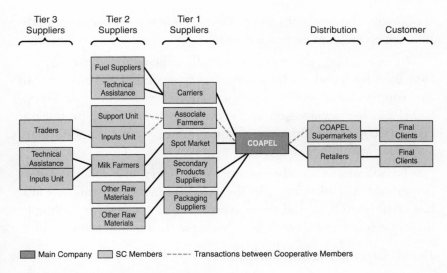

Exhibit 22-1   COAPEL Supply Chain

The co-op management is professional, but engaging in permanent interactions with the co-op's members through cooperative councils. In addition, COAPEL has a strategic guideline for management related to matters such as transparency, promotion of the co-op's actions and principles, maximization of members' benefits, and self-sufficiency, among others, based on the company's bylaws. Also, the company is mainly run according to the cooperative's principles, as evidenced in some of COAPEL's practices, such as programs and benefits offered to its members.

In this supply chain, the main transactions are the frequent interactions between COAPEL and its cooperative associates. In these transactions, there are exchanges for both sides: farmers provide raw materials, fresh fruit, and milk, while the co-op buys their production and supports them with some benefits for this production and development of their land. To better manage the range of associates, COAPEL has subdivided them into five areas. However, this management process has some limitations, since there is no strict control of production receipt. In addition, there is no association or supply contract with the co-op and its associates. However, to join the cooperative the associates accept the norms and rules of the organization, which includes certain liabilities.

In conclusion, the co-op is socially efficient with strong performance by members in management, especially the farmers. In this supply chain, process flows are controlled by the main co-op, COAPEL, but management of the chain is shared among professionals and cooperative members through councils, with large farming associates acting more actively.

These farmers are also consumers of the co-op's retail activities (supermarkets and inputs). In the retail business the company also has urban members, and these associates are managed according to their purchases, obtaining profit-sharing at the end of each year.

Focal Supply Chain Analysis

Because the main and more critical transactions in this supply chain occur between the co-op and its members, the consulting team's analysis centered on the focal supply chain presented in Exhibit 22-2.

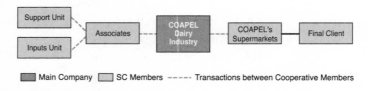

Exhibit 22-2   Focal supply chain.

Some requirements to join COAPEL include being a person dedicated to agricultural activity on self-owned land, selling all production to the co-op, and agreeing not to engage in practices that can harm or conflict with the co-op's activities. Individuals and entities not dedicated to agricultural activities, but who are interested in retail consumption, may also join.

COAPEL does not have a maximum membership limitation, but membership cannot fall below 20. The association process is completed by filling out a membership proposal—there are no formal contracts in this relationship. The entrance of a new member must be approved by the COAPEL Board of Directors, and, if there are no restrictions, the candidate will subscribe to shares and become a cooperative associate. Everybody is free to join and leave the association, and COAPEL also has the right to dismiss a member that is violating its bylaws. Associate shares are transferable to heirs, aiming to maintain farmers as co-op suppliers, but in some cases it is more financially attractive to disassociate due to the appreciation of shares. COAPEL has little control over associates leaving, while some associates remain but in an inactive state.

Members' participation occurs through the General Assembly, the main company council, where big decisions are made and the Board of Directors and Audit Committee members are elected. All members, except the ones with employment relationships, can be elected and voted into COAPEL councils. Council representatives are elected every three years, and one-third of them are allowed reelection. At least two-thirds of the Board of Directors need to be engaged in agricultural activities in order to keep focus on the co-op's core business and generate more understanding of associates' needs and feelings regarding issues such as raw material price changes. The main members of this supply chain are milk farmers. The co-op only buys raw materials directly from producers that are associates. COAPEL undertakes to buy all associates' production, and associates are not allowed to sell this production to other companies (although COAPEL has little control over this). The associates have a huge range of benefits given by the co-op, and if they divert production, the co-op can charge them the amount of investments made in their land. Most of these farmers own a small amount of land, producing less than 18,000 liters of milk every day. In particular, 55% of the farmers lie in that production range and comprise 19% of total production, 36% produce between 18,000 and 100,000 liters per day and comprise 48% of total production, and the remaining 9% produce more than 100,000 liters daily and comprise 33% of total production (COAPEL 2010).

Supply Chain Analysis Based on a Transaction Costs Approach

After the consulting team analyzed the relationships and management along COAPEL's supply chain, transactions were analyzed based on transaction costs, divided into two pieces: behavioral assumptions and dimensional assumptions.

Behavioral Assumptions (Summarized in Table 22-1)

Table 22-1   COAPEL Behavioral Assumptions

Bounded Rationality	Different cognitive capacity between big and small/medium farmers.
Opportunism	Pursuit of self-interest and economic gains by farmers.
	Use of the benefits provided by the company.
	Free membership and leave principle.
	Different treatment of members seeking economic gains for the company.
	Lack of division between ownership and control.
	Prices paid to raw material and paid by consumers in retail.

Within bounded rationality were found differences in cognitive abilities between large versus small farmers. Larger farmers have their rationality expanded, due to greater access to education and information and a better relationship with the company, community, and other farmers. These factors generate more subsidies to them in decision making and, consequently, result in further development of their properties. Small farmers tend to have less access to education and information along with restricted relationships with other members and the community, as well as range of cultural and social barriers. These cognitive capacity differences were visible in the members' respective council participation, whereby small farmers lack awareness and organizational capacity due to a bigger bounded rationality to make strategic decisions.

Members' opportunism was once more related to the farmers' respective sizes. Bigger farmers are less loyal and have weaker ties in relationships with the co-op due to both a greater independence that they have from the co-op to develop their activities and a larger focus on self-interests. This factor was evidenced in dairy farming competition, when associates choose to no longer sell to COAPEL and instead seek greater profits offered by competing companies.

This opportunism also appeared through the use of the co-op's benefits. Members can improve their lands and improve their production through subsidies and access to low-cost loans given by COAPEL and are free to join and leave the association at any time, which is a cooperative principle, without COAPEL receiving returns on those investments.

Another opportunism factor was evidenced in the dependence of small farmers with COAPEL to help make their activities economically viable. Some dissatisfaction was evident, as production bonuses are given, in general, to large farmers who have greater ability to modernize and invest in their land, impacting directly in volumes and quality of milk. The co-op tries to solve part of this problem through an active approach to small producers, seeking growth and development of their production. Besides, COAPEL faces a dilemma with this issue: it wants to maintain small farmers, even when economically unviable, while needing to retain large farmers, less representative in membership, but more economically profitable for the co-op.

As previously referenced by Batalha (2001), a lack of division between ownership and control of the company was also evident. Although the company is managed by professionals, trying to mitigate this risk, the associates actively influence COAPEL's councils, directly affecting the management. Thus, there are trends in pursuit of self-interest and seek greater benefits for farmers. Finally, there was clear evidence of huge differences between the price paid to producers and those paid by consumers.

Dimensional Assumptions (Summarized in Table 22-2)

Table 22-2 COAPEL Dimensional Assumptions

Frequency	Milk is collected every two days from farms.
	Final products are produced and distributed daily.
	Inputs are sold to farmers depending on their needs.
	Technical visits are conducted sporadically and by associate request.
Asset Specificity	Assets in the industrialized process are little adaptable to other production processes.
	Assets related to farmers' care of animals are highly specific to this activity.
Uncertainty	No contractual relations between associates and the co-op.
	Price volatility and seasonality of production, influenced by external factors.
	Investments vs. returns.

Most assets used in the dairy industrialization process are highly specific and have high monetary value. Also, equipment and machinery used for animal husbandry, as well as the animals themselves, have high value added to production and are quite specific. A dairy cow, for example, has no value or quality in its meat because its genetics result in excellent milk but poor-tasting meat.

Transaction frequency between members of the supply chain occurs on a recurring basis: (1) every two days milk is collected from the farms, (2) end products are manufactured daily, (3) these products are distributed to final clients daily, (4) product sales also occur regularly, depending on stocks and needs of consumers, and (5) technical visits occur more sporadically, depending on farmer requests.

Transaction uncertainties are also strongly related to a lack of contracts for members' milk supply. The only contracts between COAPEL and its associates are related to financing, which guarantee payments of an associate during the contract period. Thus, there is no certainty of permanence of associates or return on investments made by subsidies given.

Another uncertainty factor is related to the dairy sector that has a high volatility of prices and other external factors such seasonality of production and demand, which directly affects the supply chain activities. The co-op seeks to mitigate these risks through product reserves, maintaining stability of prices paid to its members.

Action Plans

After analyzing COAPEL's supply chain, the managers promoted a meeting with the council of cooperative members and other key staff members to brainstorm and list some possible action plans to be implemented in the co-op in order to reduce risks and improve the company process as a whole. Whatever actions emerge from this meeting cannot contradict the co-op's values and principles. Also, all the actions have to be related with each of the transaction costs that come out the analysis as a form of mitigating those problems.

References

Batalha, M.O., *Gestão Agroindustrial*. Atlas: 2001. 690p.

Chopra, S.; Meidl, P., *Gerenciamento da Cadeia de Suprimentos: Estratégia, Planejamento e Operação*. Prentice Hall: 2003. 465p.

Cooperative. In: Aliança Internacional Cooperativa – *International Co-operative Alliance*. Retrieve from <http://www.ica.coop/al-ica/>.

Cooperative. In: Organização e Sindicato das Cooperativas do Estado do Rio Grande do Sul, 2010. Retrieve from <http://www.ocergs. coop.br/>.

Cristopherson Am, G; Coath, E. *Collaboration or Control in Food Supply Chains: Who Ultimately Pays the Price?* International conference on chain and network management in agribusiness and the food industry, 2002, Holanda: Wageningen Academic Publishers.

Farina, E. M. Q. apud Zibersztajn, D., Neves, M. F., *Organização Industrial no Agribusiness*, Pioneira: 2000. pp. 39–60.

Lambert, D. M.; Cooper, M. C, *Issues in Supply Chain Management + Industrial Marketing Management*. New York: 2000. Vol. 29, pp.65-83.

Yin, R. K., *Estudo de Caso: planejamento e métodos*. Bookman: 2001. 248p.

Williamson, O. E., *The economic Institutions of Capitalism*. Free Press: 1987. 450p.

Gusmão, S. L.L., Proposição de Um Esquema Integrando a Teoria das Restrições e a Teoria dos Custos de Transação para Identificação e Análise das Restrições em Cadeia de Suprimentos: estudo de caso na cadeia de vinhos finos do Rio Grande do Sul. Ph. D. Thesis, UFRGS, Escola de Administração: 2004. 222 p.

Case 23

Continuous Process Reforms to Achieve a Hybrid Supply Chain Strategy: Focusing on the Organization in Ricoh

Mikihisa Nakano[†]

Introduction

Are reforms of business processes in a supply chain (henceforth, supply chain processes) time limited or continuous? Global companies in advanced nations that are exposed to rapidly changing conditions in both their domestic and international markets should adopt an approach for introducing continuous reforms to the supply chain. These global companies must develop supply chain processes that can simultaneously achieve two strategic targets—operational efficiency and responsiveness—to continue to mine mature domestic markets and to open up overseas markets in advanced and emerging countries. Achieving both these targets usually requires a considerable period of time.

A hybrid strategy synthesizing a lean strategy to target efficiency and an agile strategy to target responsiveness has been called lean/agility or "leagile." Because the strategy was proposed by Naylor et al. (1999), a body of theoretical and empirical study has been accumulated (Mason-Jones et al., 2000; Naim and Gosling, 2011; Qi et al., 2009; Stavrulaki and Davis, 2010). However, there has hardly been any discussion regarding what kind of organizations are best suited

[†] Kyoto Sangyo University, Kita-ku, Kyoto, Kyoto, Japan; mnakano@cc.kyoto-su.ac.jp

to this hybrid strategy. Continuous reforms in supply chain processes are essential if this challenging hybrid strategy is to be realized. What role, then, should be played by the organizations that conduct these continuous process reforms?

This case study considers Ricoh Company, Ltd. (henceforth, Ricoh), a leading Japanese company with a global presence that created a hybrid supply chain strategy for its office imaging equipment, such as copiers and printers. Ricoh has been conducting process reforms for over 10 years, resulting in a supply chain that is highly efficient and responsive. To advance these process reforms, Ricoh established a cross-functional committee to investigate and conduct reform projects and a supply chain management (SCM) promotion department to support these reform activities. In this case study, we introduce the activities undertaken by the committee and the SCM promotion department to achieve Ricoh's process reforms. With the help of this case study, the author endeavors to stimulate a debate among readers regarding why continuous process reforms are required and what organizational elements are required to make them successful.

Company Background

Ricoh manufactures and sells office imaging equipment, mainly copiers, printers, multifunctional printers, projectors, and facsimile machines. It also provides services and business solutions for these appliances. It divides its global market into the following five regions: Japan, the United States, Europe, China, and Asia Pacific. Its supervisory headquarters are in Tokyo, New Jersey, London, Shanghai, and Singapore. It provides sales and services to over 200 countries and regions worldwide, with 14 manufacturing plants in Japan and 7 overseas. In fiscal 2011, Ricoh achieved consolidated sales of 1.9 trillion yen (approximately 23.8 billion USD), and at the end of March 2012, it had 109,241 employees on a consolidated basis.

Process Reforms

The foundation of Ricoh's process reforms is a response to customers' diversified needs through a reduction in product life cycles in the second half of the 1990s that followed the shift toward digitization and networks for office imaging equipment. Previously, products would have a life cycle of two to three years; however, after Windows and the Internet became established as the de facto standards for information technology, Ricoh reduced its typical product life to slightly more than a year. Because of this change, Ricoh could no longer hold onto its surplus inventory and had a growing need to develop a supply chain that could respond rapidly to diverse customer needs at a low cost.

At the end of the 1990s, Ricoh introduced business process reforms and an integrated information system, which it used to launch reform projects aiming to achieve low-cost operations, while simultaneously increasing the levels of customer satisfaction. In the second half of 1998, Ricoh began a trial of a method that it termed "plant kitting" for its laser printer, one of its main products. Prior to this, Ricoh manufactured the main body of the various types of printers and the optional parts at its plants and these were then stored individually in sales company warehouses. Following an order from a customer, the person in charge of the sale would pull the required parts out of inventory and send them to the customer's office, where the product would be assembled and installed. Even orders that involved implementing memory boards and/or network boards, or customized tasks such as setting up an IP address, were conducted by the salesperson responsible for the order.

In contrast, plant kitting entailed a series of operations that were all conducted at the plant. First, based on the details of the order acquired from the sales company, the main body of the printer and the optional parts were manufactured separately. Next, based on the confirmed order information, the product was assembled rapidly and customized at the plant. Finally, the completed product, which was assembled according to the customer's specifications, was delivered directly to the customer. In other words, Ricoh was aiming to achieve a lean/agile hybrid strategy for its printers.

This trial reform was implemented in conjunction with several other process reforms. In 1999, Ricoh began a trial change to weekly production instead of the conventional monthly production at its plants within Japan. At the start of the trial, Ricoh established production plans for two-week periods. In 2003, Ricoh formally adopted this process using a weekly production plan, and Ricoh's plants in America and Europe also changed to weekly production.

On the product development side, in 1999, Ricoh began moving toward modularization, and in 2003, the company began to use the Modular Build & Replenishment (MB&R) production method. Under this method, modules (the core parts common to all products) are manufactured at Ricoh's consolidated production bases in Japan and China and subsequently shipped to the appropriate international production bases where the product will ultimately be used. The product is then assembled to completion based on the final specifications.

On the procurement side, in 2000, Ricoh introduced the "RaVender-Net" network infrastructure in which transactions with suppliers are conducted using electronic data interchange (EDI) to digitize quotations, drawings, orders, and other documents. In addition, Ricoh provided support for process reforms at the plants belonging to its suppliers in order to increase the accuracy of deliveries for weekly production units. On the sales side, in 2003, Ricoh's sales planning department in Japan began producing sales plans on a weekly basis. In 2005, Ricoh created a system for a weekly sales-production plan in which sales and production were linked on a per-week basis. The company subsequently expanded this approach beyond printers to its other core businesses, and in 2005, it launched plant kitting for its medium-sized multifunctional printers. In 2008, Ricoh started production based on orders for its large-sized multifunctional printers.

In conjunction with these process reforms, Ricoh also introduced a number of integrated information systems. In 1999, it introduced the Daily Production-Sales-Inventory Monitor (DPSIM), which displays data about items such as production and sales planning and results, production status, and inventory information in daily units. In 2002, it introduced Global Inventory Viewer (GIV), which displays data regarding inventory conditions, and in 2003, it introduced Production-Sales-Inventory Automation Planning System (PSIAPS), which supports sales planning.

As a result of the implementation of the reform projects described above, Ricoh's percentage of total sales from overseas grew from approximately 40% in fiscal 1999 to about 50% in fiscal 2012, whereas the percentage of total overseas production increased substantially from approximately 50% to approximately 90%. Amid the significant changes taking place in Ricoh's production and logistics networks, its inventory turnover period declined from the typical 2.6 months to less than 2 months. At the same time, it reduced the time required to deliver and install a customer's order to one-quarter of the time taken previously. It is noteworthy that Ricoh's inventory turnover period on a monthly basis has trended at a level significantly lower than that of its competitors. These results verify that Ricoh has been able to achieve both low-cost operations and increased customer satisfaction through the development of a hybrid supply chain.

Organizational Operations

The SCM group established in April 1999 was responsible for promoting this series of reform projects. The mission of this department was (1) to make business processes visible and to conduct operational reforms, (2) to make progress in companywide structural reforms by determining the framework for companywide SCM activities and promoting these activities, and (3) to optimize companywide supply chain activities through the development of a supply chain based on the customer's perspective.

After the fundamental concept for the reforms and a plan for their implementation were created within the SCM group, the Company-Wide Structural Reforms Committee (henceforth, the reforms committee) was established in February 2001. As shown in Exhibit 23-1, three types of projects were established under the reforms committee umbrella: projects to implement structural reforms (henceforth, structural reform projects), projects to develop information systems, and projects to promote SCM reforms.

Director A, who had overall responsibility for the SCM group, was appointed chairperson of the reforms committee, and the SCM group filled the staff office role. Structural reform projects were established for each department: production, product development, domestic sales, overseas sales, and services. The director in charge of each department was given the overall responsibility for the reform, whereas the department head or section head served as the chairperson of the committee for that reform. In addition, a steering committee comprising the directors in charge of each department was established. The purpose of this was to give them the opportunity to deepen their understanding of the reforms being conducted at work sites, acquire an awareness of the issues, make informed decisions, and provide advice in order to develop the members of the steering committee as general managers. Ricoh was able to create this sort of companywide structure for promoting reforms owing to the strong leadership of Director A, who was the chairperson of the reforms committee.

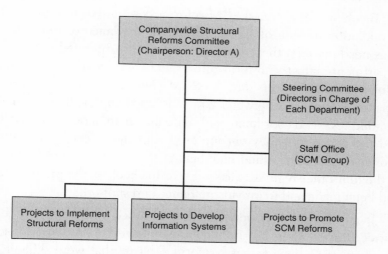

Exhibit 23-1   The committee for advancing SCM reform projects (February 2001 to November 2010).

The reform projects progressed in the following manner. In order to coordinate the reform themes, the chairpersons of the reform project committees would meet with the chairperson of the reforms committee and a representative of the staff office at the start of the fiscal year. After the reforms themes were decided upon, the members of each reform project committee would implement the reforms. The reforms committee met every month and received reports and issues presented by two or three of the reform project committees. It would share this information with, and listen to, the opinions of the members of the steering committee and the members of other project committees. In addition, in some cases, it would request the cooperation of other departments, which would subsequently be reflected in the reforms. Director A (chairperson of the reforms committee) and the

SCM group (the staff office) held regular meetings twice per month to coordinate the direction to be taken with the reforms and to share information on any issues encountered. Further, Director A and the SCM group met with the president of Ricoh once per month to report on the progress of the reforms. Both of these regular meetings were also attended by the chairpersons of the main reform project committees. When required, these chairpersons would answer questions and coordinate with members of other project committees.

Ricoh also established SCM reforms committees in four regions of operation outside of Japan. Under the guidance of the regional sales headquarters, these committees have incorporated production departments into the reform efforts and they continue to conduct reform activities to this day. An SCM general meeting was held in Japan once per year between 2001 and 2009 so that these overseas committees could coordinate and share information on their progress and the issues that they encountered with the reforms committee. Moreover, Director A and members of the SCM group visited each of the four other regions twice per year to check on the progress that had been made and to provide guidance and support.

In addition to functioning as the staff office for the reforms committee, the SCM group fulfilled the following two roles. First, it advanced cross-departmental reform projects that were difficult to categorize in each department (production and sales process reforms and logistics process reforms). Second, it advanced reforms to monitor and achieve the targets set for the two key performance indicators (KPI) of inventory and logistics cost. Specifically, it would request performance data from the relevant departments, compile all the data, prepare presentation materials, and distribute these materials to the relevant departments and top management. In addition, if performance had deteriorated, the SCM group would request a report from the relevant department outlining the causes and suggesting how to resolve these problems. On some occasions, members of the SCM group were dispatched to the relevant department to provide support. In these ways, the SCM group played an extensive role, thoroughly monitoring the progress of both cross-departmental process reforms and those being conducted by individual departments.

Suspension and Restarting of Committee Activities

The reforms committee met almost every month, for a total of 86 meetings in the 9 years from its establishment up to November 2010. However, upon the retirement of Director A, it was determined that the reform projects had reached completion, and thus the activities of the reforms committee were suspended. Nonetheless, owing to the effects of the Great East Japan Earthquake that occurred approximately half a year later in March 2011, the flooding in Thailand in the fall of the same year, and the financial crisis in Europe, Ricoh's inventory became imbalanced, with shortages of some of its best-selling products and excess inventory of products that did not sell as well. To address the changes in its operating environment, Ricoh developed an emergency supply process to deal with the problem of product shortages, to minimize inventory imbalances, and to make full use of the supply chain system that it had previously developed, including such elements as postponement, inventory visualization, and direct deliveries to customers. Despite these efforts, Ricoh was unable to stabilize its level of inventory performance, and as a result, it recorded an operating loss in the fiscal year ending in March 2012. This result demonstrated to Ricoh that it had to improve its SCM system further.

In August 2012, Ricoh established a new SCM reforms committee in Japan, influenced by the SCM group. As shown in Exhibit 23-2, the president of the sales headquarters was given overall responsibility for this committee, while the staff office was established within the sales headquarters. The same organizational structure used by the overseas SCM reforms committees was adopted in Japan. Four reform teams were established under the control of this committee: the production, sales, and inventory process reform team; the logistics process reform team; the plant kitting process reform team; and the operations reform team for information systems.

As the SCM group no longer functioned as the staff office for the SCM reforms committee, its members were incorporated into the various reform teams and became involved in the planning and implementation of the projects. Some of these members also became members of the steering committee and participated in committee decision making. Moreover, in order to continue to implement and progress companywide SCM reform projects, a team was established within the SCM group that was responsible for documenting, as intellectual assets, the expertise that Ricoh had acquired thus far by implementing its process reforms. Consequently, the SCM group became responsible for not only supporting the progress of process reforms and thoroughly monitoring them but also implementing reforms and creating intellectual assets from the expertise that Ricoh had acquired.

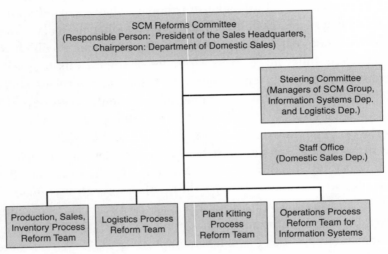

Exhibit 23-2 The committee for advancing SCM reform projects (since August 2012).

Discussion Questions

1. After Ricoh suspended its companywide structural reforms committee, the performance of its inventory worsened due to the effects of the Great East Japan Earthquake, the floods in Thailand, and the financial crisis in Europe. It has subsequently been unable to stabilize the level of its inventory performance. Can you speculate on the connection between its suspension of a cross-functional committee and the instability of its inventory performance?

2. What do you think about the organizational elements necessary to implement continuous supply chain process reforms?

References

Mason-Jones, R., Naylor, B. and Towill, D.R. (2000), "Lean, agile or leagile? Matching your supply chain to the marketplace," *International Journal of Production. Research*, Vol.38, No.17, pp.4061–4070.

Naim, M.M. and Gosling, J. (2011), "On leanness, agility and leagile supply chains," *International Journal of Production Economics*, Vol.131, pp.342–354.

Naylor, J.B., Naim, M.M. and Berry, D. (1999), "Leagility: Integrating the lean and agile manufacturing paradigms in the total supply chain," *International Journal of Production Economics*, Vol.62, pp.107–118.

Qi, Y., Boyer, K.K. and Zhao, X. (2009), "Supply chain strategy, product characteristics, and performance impact: Evidence from Chinese manufacturers," *Decision Sciences*, Vol.40, No.4, pp.667–695.

Stavrulaki, E. and Davis, M. (2010), "Aligning products with supply chain processes and strategy," *International Journal of Logistics Management*, Vol.21, No.1, pp.127–151.

Case 24

Improving Preparedness in Supply Chain Risk Management at Jacket

Jury Gualandris[†] and Matteo Kalchschmidt[‡]

Pietro Ravelli raised his head from the paper he had just finished reading. Indeed, he now perceived that his problem was not unique and that some insights could have been drawn from what has been experienced by other managers. Still, after so many years, the "risk management" issue was not completely clear to him. He understood that risk is a multidimensional concept and that it is not wise to consider remedies successful in other business contexts as universal solutions. However, he considered that good ideas could come by looking at the behavior of companies that experience supply risk under different environmental conditions. To this end, he usually spends a couple of hours more in the office to peruse recent publications covering supply chain risk management (SCRM).

After serious negative experiences, risk management had become a major buzzword in all of Ravelli's meetings. As supply chain manager of Jacket, his responsibility is to guarantee efficient flows of inbound goods. Since a couple of historical suppliers had incurred delays and quality problems, causing negative drops in Jacket's production and a temporary inability to satisfy customers' needs, Pietro's attention is now radically oriented toward the reduction of supply vulnerabilities.

Jacket, established in the mid-1960s, develops and manufactures fluid and gas control solutions (i.e., control valves and on-off valves).

[†] Università degli studi di Bergamo, Bergamo, Italy; jury.gualandris@unibg.it

[‡] Università degli studi di Bergamo, Bergamo, Italy; matteo.kalchschmidt@unibg.it

The company was founded by three small companies in 1962 that decided to share their competencies in the design and production of control valves. Started as a small corporation, in the 1980s it grew significantly, mainly through acquisition. In the mid-1990s, it was acquired by Catervision, a multinational group operating in the fluid and gas control industry that owns more than 30 companies around the world. Jacket currently hires almost 500 employees. It produces control valves in the northern part of Italy, and these are sold using the sales force of the Catervision group. In 2011, revenues exceeded $225 million with high margins.

Jacket operates in a fragmented market where business uncertainty and complexity are significant. The growing variety of market requirements and the increasing uncertainty surrounding both customer needs and the trajectory of various technologies have forced the company to operate on a make-to-order basis and arrange its business into different interrelated projects. In such a scenario, suppliers' delays or quality problems can have greater impacts on the firm's performance, both in terms of costs and reputation.

Jacket's products are managed along four product families. The procurement function manages more than 5,000 different items. These many supplies do, however, share similar characteristics in terms of material, geometry, and technology. Such goods are acquired from a few global leaders that have similar production capabilities and produce substitutable products. Supply markets are highly concentrated, and, although vendors have always shown high reliability, opportunistic behaviors are likely to occur (i.e., suppliers may collude on the price of their products or set their production capacity in a way that allows them to exploit customers' surplus). The company employs vertically specialized workers that are permanently linked to a certain production task, and it relies on an efficient production process. The production sequence was identified years ago and is still considered optimal for maintaining good manufacturing performance.

Before beginning his career at Jacket, Pietro Ravelli operated as purchasing manager in several industrial companies. In 2005, he joined Jacket as supply chain manager, and since his first year at Jacket, his attention has focused on the quality and cost of supplies. To reduce purchasing costs and optimize transportation efficiencies, he decided to establish single-sourcing relationships. These are characterized by complete, detailed contracts, with little latitude in the transaction parameters of time and cost. This practice allows Jacket to obtain quantity-based discounts. Furthermore, aiming at assuring high quality of final products, Ravelli decided to develop a vendor rating system that primarily accounts for the quality of supplies.

Nowadays, however, Ravelli is more concerned about the *reliability* of Jacket's supplies. He has recently read about four interesting and relevant cases.

Case 24-1: Helmets

The first case describes Helmets, which produces circular saw blades, cutter heads, router bits and other accessories for the woodworking market. After a major fire that shut down the operations of one of its suppliers, Helmets was able to effectively react by relying on an alternative vendor with which the company used to have spot exchanges. The company, indeed, purchases more than 80% of supplies via a multiple-sourcing strategy. At Helmets, the adoption of SCRM practices is limited: the multiple-sourcing strategy is the only lever that reduces supply chain vulnerability. The company does not have structured supply policies or procedures. Proximity and price drive supplier selection, and no information regarding vendor lead times and financial performance are collected and monitored. Furthermore, Helmets does not share key resources (e.g., knowledge, financial, and human capital) with suppliers. Instead, it simply collaborates with a small subset of them to optimize logistics processes. Internally, the firm has implemented highly automated production lines, and employees have very specific skills. The production process is highly efficient but at the same time scarcely flexible.

Helmets is located in the northern part of Italy, just 30 minutes away from Jacket's headquarters. Thus, Ravelli decided to visit the company to gather further insights. After a brief discussion with Giuppelli, the purchasing manager of Helmets, Ravelli understood that the contexts in which Jacket and Helmets operate greatly differ from a risk perspective. Helmets mainly purchases raw materials with standard quality certifications (e.g., ISO 2859 and Acceptable Quality Levels standards) from local suppliers (i.e., 60% of goods are purchased in Italy; 100% of goods are purchased in Europe). Moreover, the exogenous uncertainty of the woodworking market, as described by Giuppelli, is not critical. Customer needs are quite stable, and technology of both products and processes is consolidated (e.g., product life cycles average eight years).

Case 24-2: Puffs

The second case describes how well-known car manufacturers are managing their supply chains. By strictly monitoring their entire network and by requiring greater quality and higher levels of service, they are pushing upstream partners to increase their reliability and adopt SCRM levers extensively. The extensive adoption of such practices, however, seems to produce negative impacts on partners' economic performance. Specifically, the article focuses on Puffs, a specialized producer of thermoplastic components for electric accumulators. The company is part of a group operating worldwide, and it supplies the automotive chain. In Puffs' specific sector, market needs and technological trajectories are described as stable and predictable. Puffs manages only a few dozen different kinds of raw materials acquired from big multinational companies. As a consequence of the pressure from supply chain customers (i.e., car manufacturers), the company arranges purchases into categories by considering product and market peculiarities (i.e., level of goods' customization and complexity, market concentration, and probability of thinness) and leverages backup sources (i.e., more than 50% of purchases are managed through dual sourcing). Puffs monitors all its vendors on a monthly frequency by assessing both their operational reliability (i.e., by assessing delivery performance and also auditing safety levels and maintenance programs) and their financial stability (i.e., by evaluating past and expected cash flows).

The customer requirements have caused Puffs' internal costs to rise, but the firm has not realized positive returns from its new SCRM practices. Given Puffs' circumstances and the characteristics of its supply network, the dual sourcing and certain other SCRM directives mandated by Puffs' customers seem unnecessary. For example, Puffs is facing higher transaction costs due to the duplication of procurement processes and the higher potential for friction. At the same time, Puffs now buys at higher price, because backup suppliers have to support specific investments that increase prices, while focal firms cannot obtain discounts due to low order quantities.

Case 24-3: Sirop

The third case discusses the story of Sirop, the Italian subsidiary of a renowned international group leader in the moulding injection equipment market. The firm is well-recognized for manufacturing high-quality thermoplastic, thermosetting and elastomer injection machines, and robots; nevertheless, Sirop has experienced serious supply problems, and it has now reached the brink of bankruptcy. The environmental conditions under which the company operates are described as critical from a supply risk perspective. First, Sirop's market is fragmented and competitive. Thus, supplier delays and quality problems have instantaneously resulted in a significant drop in the company's market share. Further, customer needs have been described as heterogeneous and unstable. For instance, in just a few years, market awareness moved from price to safety to a focus on energy efficiency. This has led to numerous supply disruptions. The company has required vendors to adapt to the market shifts, but the vendors have not been able to properly react. Sirop purchases highly customized, high-tech components. This makes the cost of switching suppliers particularly high.

Despite the critical characteristics of its environment, Sirop has never paid attention to SCRM. The absence of risk management levers must be considered the primary cause of its current poor performance. For instance, Sirop does not have a formal supply strategy. It manages only sole-sourcing relationships to benefit from quantity discounts, thus disregarding the characteristics of its suppliers and increasing its exposure to supply risks (i.e., sole sourcing can imply high dependency on suppliers but lacks certain strategic benefits). Furthermore, Sirop has never reduced its adverse vendor selection risk. Structured approaches for supplier selection are not in place, and Sirop's procurement function does not consider operational and financial criteria when selecting or monitoring vendors. With regard to internal production processes, flexibility could be also improved. Although products share common modules (i.e., the company relies on modular design practices), decoupling and differentiating points along production lines cannot be easily postponed because manufacturing subprocesses are not standardized and production teams are usually linked to a specific production task and cannot be reassigned to different activities.

Case 24-4: Blower

The final case describes Blower, a medium-size company that designs and produces stationary electric driven compressors. Blower is one of the European leaders in the compressed air business and is part of a big industrial group offering products and services ranging from compressed air and gas equipment to electricity generation machines. According to Blower's CEO, the company has been able to significantly increase its performance during the last several years, both in terms of market share and cash flows.

The recent success of the company is mainly due to its agility, i.e., its ability to respond to short-term changes in demand or supply quickly, handling internal and external disruptions smoothly. First, the firm relies on backup suppliers that can provide additional productive capacity (e.g., purchases that, according to product and market characteristics, are classified as critical represent 30% of the purchasing cost and are managed via dual-sourcing and parallel-sourcing strategies). Second, Blower has been able to increase its ability to innovate by means of collaboration with suppliers. Blower's CEO has stated, "Vendor selection and development at Blower is a structured process that starts with a preliminary evaluation driven by a comprehensive set of criteria (e.g., presence of management systems such as ISO 9001 and ISO14001, assessment of suppliers' willingness to share private information, and cash flow regarding the past three years), and the process concludes with periodic audits, continuous improvement and teamwork initiatives."

Furthermore, the company is concurrently reengineering products and processes. It has broken down the production process into standard subprocesses that produce standard base units and customization subprocesses that, by adding modules, allow for fast personalization of the final product based on customer requirements. This high level of agility allows Blower to effectively respond to the significant changes that have occurred in the last decade. For instance, customers' attention has shifted from product miniaturization to energy efficiency in compressing air and gas. Both Blower and its suppliers have been able to rapidly adapt to the new requirements by leveraging shared resources and interorganizational knowledge. Furthermore, the CEO has stated that "globalization has also been contributing to increase the complexity and the uncertainty of the environment." The increased competition in the final market has led Blower to extensively rely on vendors that operate outside Europe in order to find new technologies that are not available locally (Blower purchases more than 50% of its supplies from outside the European continent). Nevertheless, the presence of detailed purchasing policies and procedures (e.g., the structured process of vendor selection and development) has helped the company in managing the enhanced complexity of the upstream network.

Jacket's Decisions

As a consequence of his reading, Ravelli has been developing knowledge and competencies about the management of supply vulnerabilities. According to his understanding, an important first step in the design of an effective SCRM strategy is the evaluation of the context in which companies operate. Once the elements that can influence both the probability and the impact of supply risks have been evaluated, the attention should focus on the configuration of the SCRM strategy. Different levers exist that can help companies secure their supply base. To arrive at an effective adoption of SCRM, Ravelli decided to re-consider all the information that he collected.

- *Helmets:* Although the extent to which the adoption of SCRM practices is scarce, this firm seems to perform well because its preparedness is coherent with the scarce riskiness characterizing its context (e.g., consolidated technologies, predictable market needs, standard supplies and low switching costs).
- *Puffs:* Exogenous turbulence and complexity are low (e.g., market needs and technological trajectories are predictable; plastic components are quite standard, and new consistent suppliers may be found rapidly). Nevertheless, the company relies

a great deal on SCRM. Overadoption of practices is producing different drawbacks and is negatively impacting economic performance (i.e., SCRM absorbs a large amount of internal resources and can increase the complexity of the supply chain by multiplying the number of supply relationships).

* *Blower and Sirop:* Ravelli perceives Sirop as risky and Blower as reliable and efficient. Both companies operate in a critical environment. The rate of product obsolescence and the frequency of technological changes are high. Make-to-order processes, foreign production plants, and many relationships with foreign suppliers lead to increased business complexity. Since purchases are complex and customized, the dependency on suppliers is high, and supply disruptions may have serious detrimental consequences on the companies' operational (e.g., lead times, sourcing, and production costs) and financial performance. Nevertheless, the level of SCRM adoption significantly differs between the two firms. Unlike Blower, Sirop defines its sourcing strategy without considering all the characteristics of its purchases. This had led Sirop toward inappropriate purchasing strategies and a serious exposure to supply risk. Further, Sirop rarely analyzes delivery capabilities and financial performance of its vendors. This is inappropriate when a large number of sole-sourcing relationships with foreign vendors have been developed. Such sourcing relationships have to be built upon a proper evaluation of suppliers; otherwise, the moral hazard risk becomes too great. Meanwhile, Blower involves suppliers in planning processes, shares information with them regarding production scheduling, and promotes the exchange of human resources. Sirop, on the other hand, has developed supply relationships characterized by high dependency without investing in suppliers' development. Finally, the two firms differ with respect to internal agility. Although both show a comparable level of product modularity, the picture is quite different when process resequencing is considered. Sirop does not have standardized production sub-processes and dynamic teaming. In Blower's case, such practices are essential when postponing operations to react to risk occurrence.

The time to develop a new SCRM strategy has arrived, and Ravelli must now apply what he has learned from his readings. As he develops his proposal for upper management, he has several important items to consider:

1. Should Jacket continue to use a single-sourcing strategy? If not, what type of arrangement would work better?

2. Should Jacket's supplier development and rating systems be modified? If so, how?

3. Given the context where it operates, is Jacket's agility high enough? If not, what might the firm change?

Case 25

Supply Chain Strategy at Zophin Pharma

Arqum Mateen[†]

Introduction

Zophin Pharma is a major player in the global generic drugs market with manufacturing facilities located primarily in India. Its biggest production facility is located in the state of Himachal Pradesh, India. Although the firm has become fairly successful in the global market, it has yet to really make a mark for itself in the domestic Indian market. Zophin has been trying to overhaul its operations so as to increase its profitability and market share in the national market.

After extensive deliberations among the senior executives of the firm, the company decided to hire external consultants to improve its performance. Navin Joseph along with his support team, who worked at the Delhi office of a major global consulting firm, started to closely work with company representatives from different hierarchy levels in the operations department. Navin and his team were experts in operations improvement tools and techniques, and they had successfully undertaken similar projects at other companies.

The Meeting

The first meeting between the external team and senior operations executives at Zophin was led by the President of Operations, Kishen Singh. During the discussions Joseph asked, "Who is your customer?"

[†] Indian Institute of Management—Calcutta, Kolkata, India; arqumm10@email.
iimcal.ac.in

The Zophin team was taken aback. Joseph continued, "It seems that for a firm of your size, you have a very small presence in the Indian market. To change that, we must connect with our customer. And, in order to really understand our customer, we must first know who our customer is." After much brainstorming, they came to the conclusion that their primary customer was actually the distributor. It was the distributor who ensured the availability of products throughout the country. In a price-driven country such as India, multiple companies sell the same drug under different names. Often, most of these are similarly priced. In such a situation, any player who is able to maximize its distribution footprint stands to gain. This is easier said than done, however, as competition is extremely intense. The market is full of international majors, domestic giants, as well as regional players. Additionally, alternative and traditional systems of medicines such as Homeopathy, Unani, as well as Siddha, which have been used in India for centuries, are also widely practiced. Moreover, due to the focus of the industry on price-based competition, maintaining profitability is also a major issue.

The discussion then moved to the requirements of the distributor. Joseph remarked, "We must identify the primary business of our customer so that we may be able to fulfill its needs better. Many retailers and distributors do not carry your product. It is essential to recognize that no matter what the other qualities of the product are, if it is not available to the customer, the company is losing money." Singh realized the importance of the discussion and said, "We never really thought of our business in these terms before." The executives continued their deliberations. After some time Ramesh Kriplani, a member of the Zophin operations team, said, "I think our customer is basically in the business of selling inventory. He essentially buys inventory, keeps inventory, and sells inventory. The more he moves his inventory, the more money he makes." Joseph and Singh agreed and then it was decided that in the light of the discussion, the key performance indicator for the distributor should be *inventory turnover*, as it will essentially determine the distributor's operating performance.

After this, they focused on the possible steps that could be taken to better serve customer needs. Joseph felt that this would require deeper probing at the operations level. It was decided that Darshan Dobriyal, a young member of Joseph's team who could understand the local dialect, would spend the next Monday in the manufacturing plant managed by Kriplani. It was the largest of the three plants located within the facility. The plant primarily produced Emzine, which was the bestselling drug of Zophin and a major contributor to its bottom line.

Plant Visit

On Monday, Dobriyal met Kriplani in his office, which was located outside the plant. There he also met Sunil Makhija and Ramesh Dongru, the senior supervisors of production and inspection, respectively. Kriplani told Dobriyal that the plant was basically divided into two zones, material production (including material dispensing that provided the raw material for starting production) and packaging. Final dispatch took place from a separate global distribution center located within the same complex. It handled final outputs from all the three production plants.

The plant was approved by the Food and Drug Administration (FDA) of the United States. It entailed stringent control and monitoring of the operating conditions as well as a stringent focus on quality across the plant. Plant personnel and visitors had to put on special clothing, and those working in the manufacturing zones had to wear masks and gloves at all times. They also had to put on special rubber boots as well as an additional layer of clothing. Dobriyal had to report back to Joseph in the evening. He put on the gear and went inside the plant with Makhija and Dongru.

Inside the Plant

The visit started with the morning plant meeting involving senior staff. The previous day's issues were discussed, and the production and dispatch plans for the day were reviewed. After that, Dobriyal decided to take a tour of the facility alone. He went inside the production modules (after putting on the required gear) and sat with the module supervisors for some time to understand the process. Later, he also watched the production process first hand. He noticed decreased levels of activity between 11:30 a.m. and 2:30 p.m. He checked the activity log books and noticed that there was a general trend of reduced activity/no activity during that period. He talked to one of the workers who responded, "Workers come in three shifts: 6 a.m.–2 p.m., 2 p.m.–10 p.m., and 10 p.m.–6 a.m. We (production and packaging staff who arrive in the early morning shift) have our lunch break from 11:30 a.m. to 12:30 p.m. Supervisory staff and other support personnel have their lunch break from 12:15 onward, and their working hours are from 9 a.m. to 5 p.m."

After having lunch inside the plant, Dobriyal sat with Dongru. During the course of the discussion, Dongru said, "Space constraints prevent all of us from having lunch together, so we have to stagger the lunch timings." Dobriyal also enquired about the overall management of the plant. Dongru said, "Ramesh follows a very hands-off approach and seldom comes inside the plant. So, we (Dongru and Makhija) handle the day-to-day functioning including micro-level planning, liasioning with the quality department, and adhering to the dispatch schedule."

The discussion continued for some time, and, after talking about some other issues, Dobriyal took his leave. As he exited the facility, his mind went to his meeting with Joseph regarding the day's findings. He started to make a mental note of all the issues that he had noticed.

Questions

1. Do you agree with the observation that the distributor is the customer of the company?

2. What alternatives to inventory turnover would you recommend for measuring the performance of Zophin vis-à-vis distributor requirements?

3. Analyze the impact of the current lunchtime arrangement at Zophin. Suggest possible avenues for improvement.

Case 26

Waste to Wealth—A Distant Dream?: Challenges in the Waste Disposal Supply Chain in Bangalore, India

M. Ramasubramaniam[†] and P. Chandiran[‡]

Introduction

Bruhat Bangalore MahanagaraPalike (BBMP), a Karnataka state government authority, had recently set an ultimatum that waste segregation at the source will be compulsory for all Bangalore citizens beginning October 1, 2012. But the mandate received a very poor response. According to BBMP data, 3,281 tons of total waste was generated that day, out of which the segregated garbage accounted for 515 tons. The response was particularly poor in southern zones.

Palike publicized in different areas of Bangalore to meet the October 1, 2012 deadline. However, the efforts did not yield the desired result because Palike failed to provide enough awareness to citizens regarding the necessity of segregation of waste at the source. There were neither door-to-door campaigns to educate the masses nor any meetings with local residents' welfare associations or voluntary organizations to create the necessary awareness. Monisha Ghoshbag, a Bangalore resident, noted:

[†] Loyola Institute for Business Administration, Chennai, India; rams@liba.edu

[‡] Loyola Institute for Business Administration, Chennai, India; chandiran@liba.edu

But the pourakarmika who comes to my street does not understand the concept of segregated waste. I have started giving her only wet waste daily. Once the dry waste accumulates I give it separately. On days when I give her both, she does not understand and ends up mixing them in her "dabba" (box). My week's efforts are wasted with her one minute's act of waste mixing. She says no one instructs her like I do in that area and yet at the end of it all, she ignores what I say.

If a tepid response to the BBMP's call is a concern on one hand, the failure of the existing landfill model is another. Even while the BBMP started experimenting with the new system, residents in villages on the city's periphery were angry because they bore the brunt of tons of unprocessed waste being piled up in the nearby landfills for years.

During the month of August, the Karnataka State Pollution Control Board (KSPCB) caved in to years of protests by residents of Mavallipura, where the city's largest landfill is located, and was forced to shut down the solid waste management site.

If the reports from KSPCB are to be believed, instead of processing the waste, the site had simply been piling on mountains of garbage without processing. This action by KSPCB also meant that the loads that were sent to other landfills that were already running well over capacity doubled and tripled overnight. These issues are only indicative of a bigger problem the city is confronting.

Bangalore: City Statistics[1]

Bangalore is spread over an area of 800 sq. km. As of 2008, the population of Bangalore was 78 lakhs (7.8 million). There are 2.5 million houses and 350,000 commercial properties in the city. According to BBMP administrative classification, Bangalore has been divided into 8 zones and further into 198 wards. The estimated municipal solid waste generation during 2009 from all sources was 3000 tons/day.

---

[1] http://218.248.45.169/download/health/swm.pdf

Out of this, the domestic waste per capita turns out to be 350 g/ day. Households alone contribute 54%, with the next major contributor being commercial establishments. Table 26-1 displays a typical composition of municipal solid waste.

Table 26-1 Typical Physical Composition of Municipal Solid Waste

Sl.no.	Material	Percentage
1	Paper	28%
2	Plastic	7%
3	Cardboard	11%
4	Textiles	5%
5	Grass/leaves/wood	5%
6	Leather	6%
7	Battery	1%
8	Electronic items	1%
9	Metal	1%
10	Organic	2%
11	Glass	24%
12	Debris	4%
13	Biomedical	5%

Collection of Waste

The system of waste collection is handled by both BBMP (35%) and contractors (65%). The contractors are selected after calling for a tender, which is basically a sealed bid lowest price auction mechanism. This process usually takes two to three months. There are about 4200 pourakarmikas (sweepers) from BBMP and 9500 pourakarmikas from contractors who perform the door-top-door collection and sweeping activities. There are some areas in Bangalore where this door-to-door collection activity is entrusted to women's self-help groups (SHGs). The members of these SHGs live below the poverty line. Even the local residential welfare associations are involved in these collection and disposal activities.

It is generally accepted that quantification of solid waste is the single most important activity in proper waste disposal. However,

BBMP does not have a mechanism with which the waste can be quantified in all of the zones.

Exhibit 26-1 displays a mapping of the waste collection supply chain.

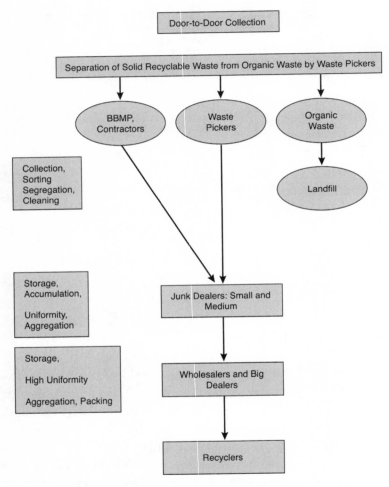

Exhibit 26-1   Waste collection: a supply chain mapping.

Door-to-Door Collection

Door-to-door collection is the primary means of waste collection in Bangalore, followed by the community bin method. In the community bin method, waste is stored in large, concrete circular bins. There are a lot of concerns raised for the community bin storage. For instance, bin designs are standardized and in many places are not proportional to the amount of waste generated.

In door-to-door collection, pushcarts and auto tippers are used for collecting the waste from each household. There are approximately 10,000 pushcarts and 600 auto tippers available for the door-to-door collection. It should be observed that until September 30, 2012, the waste was collected in an unsegregated format.

Secondary Collection and Transportation

The waste collected by the BBMP and contractors is brought to a common aggregation point. Vehicles such as compactors, tipper lorries, dumper placers, and mechanical sweepers are used by both BBMP and contractors. At this point, many waste pickers collect solid materials, such as plastic, polythene bags, papers, card boards, and PET bottles from the dumped garbage. Waste pickers sell these materials to junk dealers to earn money. Some of the contract workers are also involved in this commercial activity. From here, the waste is transported to the decentralized processing plants, dry waste collection centers, and landfills.

Recycling

Waste pickers form part of larger network of waste recycling in the country. They carry out the valuable economic activity of collecting different kinds of solid waste from streets and dumping grounds, cleaning/sorting/segregating them into different materials, and selling these to junk dealers. Waste pickers collect all solid materials from municipal waste wherever it is available. They collect up to 10 to 15 kgs. of each material per day. Materials collected at the waste picker level are classified into different qualities or grades while moving up the recycling supply chain.

Most of the value-added activities such as segregation, breaking materials, and some cleaning are done by waste pickers. But they are the lowest-paid members of the chain. The working conditions of these waste pickers are appalling. They do not have sufficient protection and are sometimes exposed to hazardous waste. A study carried out in 2003 revealed that most of the rag pickers have some kind of a respiratory disease.

Subsequent to the primary step in the recycling process, materials are sorted into low-value, medium-value, and high-value material when they are bought by junk dealers and wholesalers, and prices also vary accordingly. Junk dealers add value to material by means of better storage, accumulation, and some sorting. Wholesalers add value to material by converting waste material into a form that can be used by recyclers. Sometimes the number of stages in the chain increases and hence the efficiency decreases. For example, a small junk dealer may sell all his material to a large junk dealer. In this case, the number of stages increases to five from four. With more intermediate stages, the total supply chain cost increases.

Materials such as cardboards are sold to factories in the nearby state, TamilNadu, which uses them for printing newspapers. Wholesalers normally deal with few materials such as paper and cardboard or plastic. Even metals are traded by wholesalers to industry.

In addition to composting methods, technologies such as anaerobic digesters and incineration are used as a means of disposal of organic waste. However, these methods have not taken off successfully. For instance, although composting is a popular method, it is not economically viable because of the required maintenance. Successful composting requires continuous feeding of the organic waste, and it has been difficult to sustain the feeding. Anaerobic digesters have appeared on a small scale, but there are not many large-scale digesters in all of India.

Landfilling

Prior to closing the Mavallipura site, BBMP, using a public-private-partnership model, had been operating four landfill waste disposal sites. Various technologies are used for processing waste at landfills such as vermi composting and biomethanization. Special mention has to be made of uncontrolled landfilling, because it is the most popular method of waste disposal in India. There are systematic ways in which waste could be dumped, but to save money these are not being followed at the dumping sites.

Role of Non-Governmental Organizations (NGOs)

NGOs fill an important void in the waste management process. A variety of activities are carried out by NGOs that are instrumental in efficient functioning of the system:

- Public grievances are collected during the meetings organized by NGOs, and the complaints received are routed to the proper authorities for suitable action.
- They are involved in helping local authorities with the primary collection of waste.
- Training and awareness programs for waste pickers are conducted by NGOs.
- Some NGOs are also engaged in setting up and operating decentralized waste processing plants.

Conclusion

The waste management process is a lengthy and complex process that is the primary responsibility of BBMP and its associated contractors. This process is aided by waste pickers who form an important entity in the system. They not only collect waste, but they also perform the important activity of sorting and segregation before selling the materials. However, they are the lowest-paid and most-neglected entities in the process. The current system of waste management by BBMP has failed to efficiently implement and manage the entire process. BBMP and the contractors focus only on economic benefits and do not look at the social costs. Also, the recent move to adopt source segregation has faulted primarily because of failure on the part of BBMP to create necessary awareness in society. Such issues can only be addressed if one thinks beyond the economic benefits for the waste management operators.

Discussion Questions

1. What are the drivers of the waste management problems?
2. What are the various roles of stakeholders in this waste recycle supply chain?
3. Can this supply chain be improved, and through this can all benefit?
4. How should waste pickers be integrated with the system, particularly if private corporations get involved?
5. How can NGOs help identify solutions to these problems?
6. What steps can BBMP take to improve the current system?

Case 27 ⎯⎯⎯⎯⎯

Transitioning the Supply Network of Chennai Engineering Ltd to Cloud Computing[1]

N. Chandrasekaran[†]

About the Company

Chennai Engineering Ltd (CEL) is a large engineering company of over 45 years of existence. The company has been involved in power sectors producing auxiliary equipment and now has become a leading power sector player in India. Chennai collaborates with capital equipment players from Germany, Japan, and India to supply "balance of plants" (BOP) systems and equipment to power projects. Appendix 27-1 provides details of CEL's balance of plants scope. The company is more than Rs.4000 crores in size and does a number of turnkey projects in the midsize power sector and for corporations that want to have in-house captive power generation for manufacturing.

The Power Sector in India

The power sector has received a major thrust in power generation in recent years (see Appendix 27-2). There are a number of

---

[1] The views expressed here are of individual capacity and to be used strictly for academic discussions only. The facts, figures, and financials are illustrative and not factual or observed in any company. Neither the author nor the company is responsible for any commercial decision based on this case material.

[†] Take Solutions, Ltd. & Loyola Institute for Business Administration, Chennai, India; nchandrasekaran@takesolutions.com

challenges that lie ahead for the economy as the demand and supply gap increases. Project delays are not only expensive from the vendor side with penalties, but they are also a phenomenal loss to society in terms of socioeconomic costs.

The power sector is ranked sixth among the leading players of the Indian economy. The sector has attracted $4.6 billion in foreign direct investment between 2000 and 2012. The power ministry is believed to have set a target for adding 76,000 MW of electricity capacity in the 12th Five-Year Plan (2012–17) and 93,000 MW in the 13th Five-Year Plan (2017–22).

There are innumerable challenges that result from the gaps that exist between what is planned versus what the power sector has been able to deliver. One of the reasons for the gaps is delay in the execution of BOP jobs, which is primarily due to the absence of competent BOP players that can execute the job within stipulated time and cost.

The Challenge

Chennai's CEO, Mr. Suresh Ram, held a meeting with the CIO, Mr. Ravi Khanna, and the Vice President—Supply Chain, Ms. Deepa Pillai, for a discussion on technology deployed for the supplier network and how it could introduce more efficiency. He briefed them about the dinner meeting he attended the previous evening wherein he was invited as a guest for the launch of cloud-based solutions for the supply chain domain. The solutions are launched by a Chennai-based supply chain products company named Take Solutions Ltd, along with HP. Suresh told his colleagues that he had a decent interaction with other CEOs at the function who were of the opinion that cloud would be the way forward. If smart phones can catch up or ERP for SMEs can be deployed, Suresh tended to believe that cloud-based IT for infrastructure, business processes, and enterprise application could be important. Also, he opined that since they are located in a cluster of engineering firms where 90% of their suppliers are located, a cloud-based cluster community could bring efficiency to the supply chain.

Suresh suggested that Ravi and Deepa must explore and submit a plan on how to approach getting CEL's IT and its supplier network on cloud-based IT, especially for supply chain management. He wanted a road map and asked them to clearly articulate the benefits for CEL and its suppliers and how the solution could improve overall supply chain goals of cost efficiency and responsiveness. Deepa and Ravi responded that they would present to him in a week's time after exploring the details.

Cloud Computing

Deepa suggested to Ravi that they must spend some time understanding cloud-based solutions and its benefits. Ravi described the following key points about cloud computing:

1. An end-to-end cloud service offering can be designed using a combination of any of the four layers consisting of Software as a Service (SaaS), Platform as a Service (PaasP, Infrastructure as a Service (IaaS), and Desktop as a Service (DaaS), which is the cloud clients.

2. Cloud could be deployed through

   a. *Public cloud,* where storage and other resources are made available to the general public by a service provider. Generally, public cloud service providers such as Amazon Web Services (AWS), Microsoft, and Google own and operate the infrastructure and offer access only via the Internet.

   b. *Community cloud,* which shares infrastructure between several organizations from a specific community with common concerns (security, compliance, jurisdiction, etc.). This can be managed internally or by a third party and hosted internally or externally. The costs are spread over fewer users than with a public cloud.

   c. *Private cloud* is cloud infrastructure operated solely for a single organization, whether managed internally or by a third party and hosted internally or externally. Undertaking a private cloud project requires a significant level and degree of engagement to virtualize the business environment. Private clouds have attracted criticism because users still have to buy, build, and manage them and thus do not benefit from less hands-on management.

d. *Hybrid cloud* is a composition of two or more clouds (private, community, or public) that remain unique entities but are bound together, offering the benefits of multiple deployment models. Hybrid clouds provide the flexibility of in-house applications with the fault tolerance and scalability of cloud-based services.

3. An IDC study commissioned by AWS found that the five-year total cost of ownership (TCO) of developing, deploying, and managing critical applications on AWS represented a 70% savings compared with deploying the same resources on-premises or in hosted environments. IDC findings show that the average five-year ROI of using AWS is 626%. Furthermore, over a five-year period, each company saw cumulative savings of $2.5 million per application. And TCS savings included savings in development and deployment costs, which were reduced by 80%. In addition, application-management costs were reduced by 52%, infrastructure support costs were cut by 56%, and organizations were able to replace $1.6 million in infrastructure costs with $302,000 in AWS costs. The IDC study also showed that benefits increased over time. The study found a definite correlation between the length of time customers have been using AWS services and their returns. At 36 months, the organizations are realizing $3.50 in benefits for every $1 invested in AWS. At 60 months, they are realizing $8.40 for every $1 invested.

4. When calculating ROI in a cloud offering, one must note that the IT capacity and IT utilization would have a significant impact on infrastructure. IT capacity is measured by storage, CPU cycles, and network bandwidth or workload memory capacity as indicators of performance. IT utilization is measured by uptime availability and volume of usage as indicators of activity and usability. Other benefits that determine ROI on cloud offerings include the following.

   a. *The speed and rate of change:* Cost reduction and cost of adoption/de-adoption is faster in the cloud. Cloud computing creates cost benefits by reducing delays in decision costs by adopting prebuilt services and a faster rate of transition to new capabilities. This is a common goal for business improvement programs that are lacking resources and skills and that are time-sensitive.

b. *Total cost of ownership optimization:* Users can select, design, configure, and run infrastructure and applications that are best suited for business needs.

c. *Rapid provisioning:* Resources are scaled up and down to follow business activity as it expands and grows. They are redirected.

d. *Increased margin and cost control:* Revenue growth and cost control opportunities allow companies to pursue new customers and markets for business growth and service improvement.

e. *Dynamic usage:* Elastic provisioning and service management targets real end users and real business needs for functionality as the scope of users and services evolve seeking new solutions.

f. *Enhanced capacity utilization:* IT avoids over- and under-provisioning of IT services to improve smarter business services. Capacity additions can be staggered and spread among users.

g. *Access to business skills and capability improvements:* Cloud computing enables access to new skills and solutions through cloud sourcing on-demand solutions.

These measures define a new set of business indicators that can be used to create a "score card" of an organization's current and future operational business and IT service needs related to cloud computing potential.

Ravi stated that although the cloud has a numerous benefits, there are certain challenges for an ecosystem when it is to be implemented. Deepa suggested that the challenges can be reviewed after they discuss current IT applications in their supply chain domain and then look at the feasibility of supply chain business process application in the cloud.

Supply Chain Structure of CEL

Exhibit 27-1 displays CEL's supply chain network. Important issues include the following:

- The power project promoter identifies the project scope and engages an EPC contractor for the project, either internally or from the market.

- The EPC contractor places power island equipment orders based on the engineering services recommendation.

- The power island scope also provides scope for BOP. There would be a few vendors who would be handling BOP along with CEL.

- CEL will work with power island suppliers for information exchange and supply its plant as directed and approved by the EPC contractor.

- BOP will have its own manufacturing units and will work with a number of fabrication and other ancillary support units for completing the project items and supplying to sites where its supply chain depends on sequencing and completion by other players.

- BOP will have a number of material and services suppliers who could be also operating with competition as well as with power island equipment suppliers.

- Information exchange and completion of project components on time is critical. Typically, these projects have huge penalties for delays and failures.

In 2008, Deepa was convinced that there are a number of important supply chain management issues. CEL incurs a supply chain cost of about 12% of its cost of goods sold. The supply chain suffers from a number of problems, including (1) long procurement cycle times, (2) material delivery delays from suppliers leading to penalty costs for the supplier and for the system as the subsequent material could be ready but is held up in the system, (3) late payments to suppliers, (4) unfavorable measures for Days Purchases Outstanding, (5) manual processing of transactions on both sides, (5) lack of end-to-end control and visibility, and (6) high cost of processing transactions.

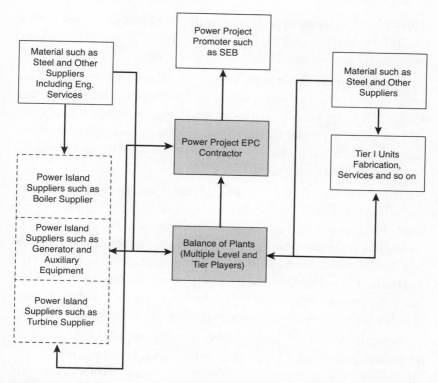

Exhibit 27-1   Supply chain network of power equipment plant.

IT in SCM

In 2009, Deepa implemented an ERP-integrated supplier network system suggested by a leading supply chain solutions company. Prior to this, CEL had handled supplier relations management via email. Appendix 27-3 shows how it was operating earlier and the improvements in processes that were brought in. The project was implemented in 90 days, wherein more than 750 suppliers automated their sourcing processes with CEL.

Benefits of the supplier integration included (1) reduction in procurement cycle time by an average of 12 days; (2) RFQs converted into purchase orders and issued on the same day for many jobs; (3) AP invoices matching reduced by 50%, which made vendors happy; (4) elimination of redundant processes and data entry, improving process efficiency; (5) only exceptions in procurement processes handled via email alerts, and phone follow-up reduced by 90%; (6) delays reduced significantly and resolved with the project sponsor.

Overall, supply chain costs fell by 8%. The number of follow-up meetings and activities was reduced, producing a further hidden cost savings. The supply chain cost savings of approximately Rs. 20 cores paid back the investment in less than six months. Deepa believed that the year-to-year maintenance expenses of Rs. 2 cores were insignificant for this size of operation.

However, Ravi had a different perspective. He was of the view that all savings could not be directly attributed to technology investment. He also felt that although Rs. 8 crores was the cost of the IT project, there were a number of management change costs that could be estimated to be as high as Rs. 20 crores. The discussions did not end there. Every function and its activities were to be streamlined and claim the benefit of change. The challenge was not just estimating benefits from business efficiencies, but something beyond.

Nevertheless, they both agreed that the supply chain business process automation has facilitated (1) increased revenue for CEL and suppliers as the transparency helps to eliminate redundancies and blocking of capacity, (2) improved financial flows that help suppliers commit to work schedules, (3) improved supply chain risk management that increases the chances of no penalty payments, and (4) increased business for suppliers as the network is exposed to better information about their respective offerings and capabilities.

Ravi also concurred that there were additional benefits when they implemented the sourcing procurement suite. These included (1) no platform investment cost for additional participants, (2) minimal business process change enforced, (3) minimal internal resource requirements for changeover, and (4) ease of training and adoption.

Discussion Questions

Deepa is pondering whether to probe further on cloud computing for the supply chain network. Would that preclude CEL having to move on bringing the supply network into the cloud? Would it be expensive to implement? Would there be a need to convince the suppliers to switch to cloud? This may not involve certain additional costs as they have already been operating integrated systems. However, two challenges may include the following:

1. If the suppliers have to pay per use for subscription, how much could they afford?

2. After the switch to cloud-based solutions, if they are also working with other vendors and power island players, how would they handle the technology of different customers? There could be a lot of cluster-driven benefits. How could one push this cluster ahead of others in technology adoption?

References

"ERP – A manufacturing perspective," www.leon-leon.com/it/erpdse/downloads/appendixc.pdf

Ferguson, Donald F. and Ethan Hadar, "Optimizing the IT business supply chain utilizing cloud computing," http://www.isaca.org/Groups/Professional-English/cloud-computing/GroupDocuments/Optimizing%20The%20IT%20Business%20Supply%20Chain%20Using%20Cloud%20Computing.pdf

http://mobile.eweek.com/c/a/Cloud-Computing/Amazon-Web-Services-Use-Yields-626-ROI-More-IDC-748758/

http://www.networkworld.com/news/2012/080612-cloud-roi-261431.html

https://www.roccloud.com/roi

www.takesolutions.com

Energy Statistics 2012, Central Statistical Office, NSO, www.mospi.gov.in

The author acknowledges the support of Mr. S. Sridharan, Mr. Ramesh G, and Mr. Sai Sridhar H for their comments.

Appendix 27-1: CEL's Balance of Plants Business in the Power Sector

The BOP or Balance of Plants system comprises all of the systems and utilities that are required to run thermal plants from raw material input to waste output, apart from the power island that includes the generator, turbine, and boiler with its auxiliaries. CEL's offering also includes the system interface engineering between various packages and between main plant and BOP packages.

BOP Unit	Remarks for a Typical 500 MW Power Plant
Coal/lignite handling plant	Value Rs. 80 crores; About 10+ component suppliers
Ash handling plant	Rs. 90 crores; 5 to 6 component suppliers
Cooling tower	Rs. 100 crores; About 3 component suppliers
Water system	Rs. 15 crores; About 10 component suppliers
Chimney	Rs. 40 crores; About 3 component suppliers
Fuel oil handling & storage system	Rs. 7 crores; About 4 component suppliers
Compressed air system	Rs. 3 crores; About 3 component suppliers
Mill reject handling system	Rs. 3 crores; About 3 component suppliers
Heating, ventilation, and air conditioning	Rs. 6 crores; About 4 component suppliers
Fire protection system	Rs. 15 crores; About 3 component suppliers
Switchyard & electrical system for BOP and so on	Rs. 70 crores; About 20-24 component suppliers

Notes: The Remarks column displays both the value and number of component suppliers. Also, it should be noted that there could be parallel projects and supplier-customer relationships running simultaneously. More importantly, for each order, a number of discussions, amendments, and transactions guidance occurs. Whenever there is a variance, it is followed with a number of touch points. Since these are engineered-to-order or make-to-order kinds of projects, the transaction touch points become critical.

Appendix 27-2: Trends in Installed Generating Capacity of Electricity Nonutilities in India from 1970–71 to 2010–11

As on	Utilities					Non-utilities		Total	Energy	
	Thermal°	Hydro	Nuclear	Total	Railways	Self-Generating Industries°°	Total	Total	Consumption in Billion kwh	Per Capita Energy Consp. (kwh)
31.3.1971	7,906	6,383	420	14,709	45	1,517	1,562	16,271	663.99	1204.39
31.3.1976	11,013	8,464	640	20,117	61	2,071	2,132	22,249	840.53	1361.74
31.3.1981	17,563	11,791	860	30,214	60	3,041	3,101	33,315	1012.58	1471.09
31.3.1986	29,967	15,472	1,330	46,769	85	5,419	5,504	52,273	1477.50	1928.51
31.3.1991	45,768	18,753	1,565	66,086	111	8,502	8,613	74,699	1902.75	2232.5
31.3.1996	60,083	20,986	2,225	83,294	158	11,629	11,787	95,081	2436.77	2593.58
31.3.2001	73,613	25,153	2,860	101,626	-	16,157	16,157	117,783	3154.28	3047.81
31.3.2006	88,601	32,326	3,360	124,287	-	21,468	21,468	145,755	3909.37	3497.59

| As on | Utilities | | | | | Non-utilities | | Total | Energy | |
	Thermal°	Hydro	Nuclear	Total	Railways	Self-Generating Industries°°	Total		Consumption in Billion kwh	Per Capita Energy Consp. (kwh)
31.3.2007	93,775	34,654	3,900	132,329	-	22,335	22,335	154,664	4226.78	3727.24
31.3.2008	103,032	35,909	4,120	143,061	-	24,986	24,986	168,047	4508.26	3928.16
31.3.2009	106,968	36,878	4,120	147,966	-	26,980	26,980	174,946	4845.25	4171.56
31.3.2010	117,975	36,863	4,560	159,398	-	28,474	28,474	187,872	5462.31	4646.87
31.3.2011 (p)	131,279	37,567	4,780	173,626	-	32,900	32,900	206,526	5693.54	4816.44
Growth Rate of 2010-11 over 2009-10 (%)										
	11.28	1.91	4.82	8.93		15.54	15.54	9.93	4.23	3.65
CAGR 1970-71 to 2010-11 (%)										
	7.09	4.42	6.11	6.21		7.79	7.72	6.39	5.38	3.44

°From 1995-96 onward, Thermal includes Renewable Energy Resources.

°°Capacity with respect to Self-Generating Industries includes units of capacity one MW above.

CAGR: Compound Annual Growth Rate = [(Current Value / Base Value)^(1 / (number of years − 1))]×100

Source: Central Electric Authority.

Appendix 27-3: Business Process of CEL Sourcing

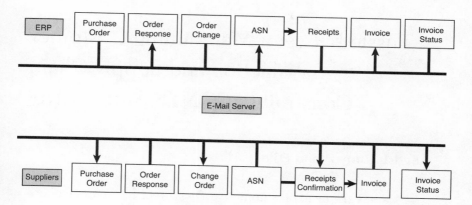

Exhibit 27-2   Before SCM Suite implementation.

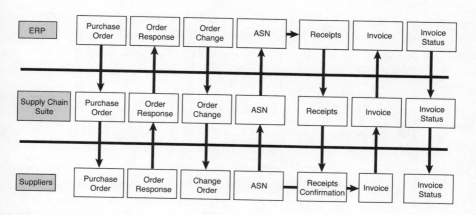

Exhibit 27-3   After SCM Suite implementation.

Case 28

Tussle between Maintaining Customer Satisfaction and Supply Chain Constraints: IGNYS Automotive[1]

Satish Kumar[†] **and Dileep More**[‡]

Mr. Neil Parr, head of Automotive Division of IGNYS, is facing a tough time. He needs to prepare a presentation to the board of directors of the IGNYS Company regarding the dip in customer satisfaction that was highlighted in yesterday's JD Power and Associates market research report. Ms. Arunima Pandey, the marketing and customer care director, and Mr. Rajesh Kumar, operations head of IGNYS, are helping Mr. Parr to prepare for the presentation. Ms. Pandey is worried about the way supply chain efficiency of the spare parts operation is deteriorating day by day. She is also concerned about the measures taken by Mr. Kumar to improve the existing situation.

The marketing and customer care unit mainly interacts with the dealers in the downstream supply chain. As an improvement initiative, the unit has identified dealers who are placing maximum orders and given special consideration to apply inventory management at their end. The unit has also implemented a dealers' engagement program through which they are making dealers more energetic and engaged

[1] As this is a pedagogical case, certain data have been disguised to ensure data confidentiality. None should be interpreted as being the actual data from this company.

[†] Indian Institute of Management, Calcutta, India; satishk2013@iimcal.ac.in

[‡] Indian Institute of Management, Calcutta, India; dileep_more@iimcal.ac.in

with IGNYS automotive. However, for the unit managing 36,143 automotive parts of vehicles and looking after 4,325 active parts, dealing with 2,316 vendors and 190 dealers is indeed a difficult task.

After discussing with the senior management, Mr. Kumar has appointed JP & Company, a top consulting firm, to help IGNYS identify the constraints in, and optimize the performance of, the supply chain for the spare parts business. After interviewing various stakeholders and via data crunching, JP & Company has given its suggestions. Ms. Arunima Pandey is still worried about the implementation component of the suggestions. Meanwhile, Mr. Kumar has tried to implement a few of the consulting firm's suggestions; however, the supply chain gap is widening daily, and management is worried about it.

Company Background

In 1986, the IGNYS Group was set up in Gujrat, India, in the manufacturing automotive segment of light and medium commercial vehicles. Today, IGNYS is in the top 20 in the utility vehicle segment in India with its flagship brands in the automotive businesses. Over the past few years, IGNYS has expanded into new industries and geographies entering into the two-wheeler segment and opening plants outside of India in order to increase its global footprint.

Today, IGNYS is public company with headquarters in Bangalore, India, and having a turnover of $10.13 billion. In total, 12,568 employees are serving the organization in its Indian and global operations. There are 10 stock keeping units (SKUs) in the company product portfolio, including SKUs present in small-segment cars, medium-segment multipurpose vehicles, midsize pickup trucks, and heavy-duty trucks. The company has 190 dealerships across 28 states, and the supply chain is being served by 2,316 vendors primarily concentrated in the four industrial hubs of Chennai, Pune, Gurgaon, and Hyderabad. The company is doing well as far as sales are concerned in India, and its branding and advertising is excellent with low after-sales service cost for consumers. However, there is also intense competition in this market from domestic and foreign players, and spare parts management is becoming tougher due to presence of non-genuine spare parts in the market.

There are many non-original equipment manufacturers that supply spare parts in the spare parts market, and they are competing with IGNYS brands on price. However, Standard Motors and India Wheels are the strongest competitors of the IGNYS. Standard Motors offers high margins to vendors and distributors on spare parts, but it also imposes stiff penalties in the case of untimely deliveries from vendors and a high number of urgent orders from distributors. On the other hand, India Wheels has bargaining power strength since it has the largest market share in the industry. This firm forces its vendors to deliver parts on time and levies heavy penalties on distributors in case of highly urgent orders from them. Compared to Standard Motors and India Wheels, IGNYS offers relatively higher margins to the vendors and dealers; however, it has low bargaining power with its vendors due to its low market share in the industry.

Spare Parts Supply Chain Management at IGNYS Automotive

Whenever the customer experiences any fault in the vehicle, an order is placed by the customer through the dealer. Depending on the urgency of the order, the dealer places the order either in *urgent* order category or the *relax* order category. The serving time for the urgent orders is two working days and for relax orders is 10 working days as promised by IGNYS. As expected, the margin for dealers is much higher in relax orders. Serving time includes order confirmation and order dispatch.

The dealer places the orders through SAP data base management software, and the orders are routed to the mother warehouse. IGNYS currently has three warehouses located at one place in the country. In the urgent order category, every day an SAP run takes place, and, if the material is present in the warehouse, it will automatically be assigned to this order. If the material is not present, a notice is sent to the respective sourcing manager of the spare parts division, who places the order with the vendor. The vendor delivers the material to mother warehouse, and then it is shipped to the respective dealer by third-party logistics. For relax orders, the SAP scanning is done every day for one region (the country is divided into the 6 regions). If the material is present in the warehouse, it is allocated to the orders; otherwise, the same procedure (as for urgent orders) applies.

In spare parts management, the Carry Forward Agents (CFAs) play a crucial role in the supply chain as they act as small company warehouses close to the customer end. They are responsible for the highly frequent relax orders so that those orders can be directly served from them. There are currently 10 CFAs of IGNYS in all of India's big states as shown in Exhibit 28-1. However, due to small size, CFAs are not able to serve their core purpose effectively. The other major problem the CFA faces is idle inventory, which is consuming a substantial part of the total space of the CFA, as shown in Table 28-1. The table shows that a large amount of inventory for different SKUs is idle for long time, which ties cash in the inventory and occupies storage space of CFAs.

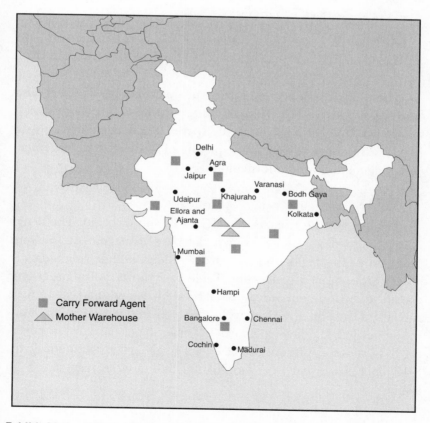

Exhibit 28-1   CFA and Mother Warehouse locations of IGNYS in India.

Table 28-1 CFA's Idle Inventory Data

No. of Days (Idle)	Value of Stock (Rs)	No. of Days (Idle)	Value of Stock (Rs)	No of Days (Idle)	Value of Stock (Rs)
786(SKU$_1$)	300	493(SKU$_{15}$)	50,000	308(SKU$_{27}$)	4,000
644(SKU$_{28}$)	52,000	491(SKU$_{11}$)	100	307(SKU$_{35}$)	4,000
617(SKU$_{18}$)	3,000	459(SKU$_{19}$)	3,000	276(SKU$_4$)	18,465
615(SKU$_6$)	1,000	430(SKU$_{10}$)	11,000	256(SKU$_{20}$)	45,768
610(SKU$_{12}$)	40,000	398(SKU$_{26}$)	4,000	253(SKU$_{22}$)	2,389
583(SKU$_{24}$)	500	386(SKU$_{31}$)	50,000	245(SKU$_{14}$)	13,456
557(SKU$_{33}$)	5,000	378(SKU$_7$)	3,560	227(SKU$_3$)	1,000
549(SKU$_{13}$)	30,000	367(SKU$_{21}$)	600	214(SKU$_{23}$)	3,000
521(SKU$_{29}$)	900	353(SKU$_{16}$)	17,839	199(SKU$_{30}$)	3,876
520(SKU$_8$)	50	346(SKU$_9$)	2,367	189(SKU$_{36}$)	500
518(SKU$_{17}$)	20,000	337(SKU$_{32}$)	3,000	184(SKU$_5$)	5,000
511(SKU$_{34}$)	4,000	322(SKU$_{38}$)	26,000	171(SKU$_2$)	6,000
499(SKU$_{37}$)	8,000	317(SKU$_{25}$)	45,664	168(SKU$_{39}$)	12,367

There are a number of players such as vendors, CFAs, dealers, customers, and the IGNYS main and spare parts sourcing divisions themselves that are involved in the spare parts management system. Exhibit 28-2 shows the complete picture of spare parts management, including material and information flows.

Dealers

Dealers are the front face of IGNYS's supply chain. The dealer places spare part orders through database management software. Many times, the dealer uses its own discretion whether it should place a relax order or an urgent order. Table 28-2 shows typical order type classification of urgent orders for an SKU by dealers, and the trend is similar for the other SKUs as well. The high-frequency O1, medium-frequency O2, and low-frequency O3 order types have orders covering 8–12 months (presence of at least a single order continuously for 8-12 months in a year), 4–7 months, and 1–3 months, respectively, in a year.

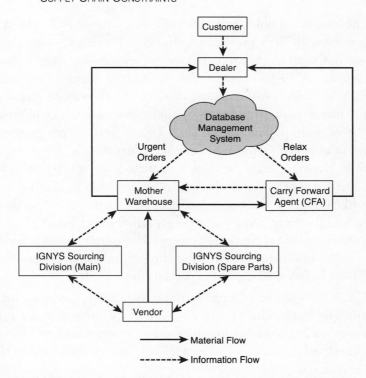

Exhibit 28-2 Process map of the IGNYS spare parts supply chain.

Table 28-2 Order Type Classification for Urgent Orders

Month	Order Type			
	O₁	O₂	O₃	Total Urgent Orders
Jan, 2012	43%	7%	50%	5,263
Feb, 2012	42%	7%	51%	5,678
Mar, 2012	43%	4%	53%	5,643
Apr, 2012	39%	3%	58%	4,567
May, 2012	42%	5%	53%	7,865
Jun, 2012	38%	4%	58%	5,656
Jul, 2012	45%	6%	49%	6,754
Aug, 2012	46%	8%	46%	8,765
Sep, 2012	48%	9%	43%	5,875
Oct, 2012	43%	6%	51%	4,567
Nov, 2012	37%	5%	58%	5,444
Dec, 2012	42%	7%	51%	5,555

The dealers should maintain sufficient inventory levels of the spare parts with high purchase frequency. However, many dealers who do not want to tie up their money in inventory usually place a high number of urgent orders. This is creating difficulty for IGNYS to fill those orders. After visiting dealers, the consultant has identified the major constraint of the situation as non-adherence of on-time delivery of spare parts to the dealers. The dealers are not getting the urgent orders in two days, and the relax orders usually take more than a month. Hence, even if the dealer wants to place a relax order he is forced to put that order in the urgent order cycle so that he would more likely receive the order in 5–10 days. Sometimes the dealers have a bad payment history and poor performance; hence, IGNYS's warehouse may not serve these dealers if they order through the relax order route. However, if they order through urgent order route, the warehouse has to send the spare parts to these dealers.

The high number of urgent orders is creating pressure on the warehousing operations. Table 28-3 shows the urgent order fulfillment rates in 2012, where, in January, for example, 23% of all orders were filled within two days, 40% of all orders were filled within three days, and so on.

Another concern is that genuine spare part prices are high relative to market prices of non-genuine spare parts. Due to this trend, IGNYS is losing market share. Yet another concern is tracking of the spare parts orders. It is difficult because the third party is often not able to trace its trucks due to the unavailability of mobile signals at various parts in India. Finally, many dealers have not implemented proper inventory management practices and hence ask for too many regular orders of which otherwise they could have had sufficient stock.

Table 28-3 Cumulative Urgent Order Fulfillment Rates in 2012

Month	0-2	3	4	5	6	7	8	9	>9
					No. of days				
Jan, 2012	23%	40%	50%	55%	60%	65%	65%	65%	100%
Feb, 2012	34%	42%	44%	53%	55%	60%	65%	70%	100%
Mar, 2012	20%	25%	35%	45%	50%	57%	63%	68%	100%
Apr, 2012	22%	32%	37%	55%	60%	66%	69%	73%	100%
May, 2012	12%	24%	36%	48%	60%	72%	73%	74%	100%
Jun, 2012	33%	33%	50%	57%	65%	67%	67%	69%	100%
Jul, 2012	23%	41%	52%	61%	67%	70%	70%	72%	100%
Aug, 2012	23%	33%	44%	55%	66%	77%	77%	80%	100%
Sep, 2012	12%	21%	31%	43%	54%	57%	60%	78%	100%
Oct, 2012	23%	32%	43%	57%	61%	64%	64%	65%	100%
Nov, 2012	25%	35%	45%	56%	60%	67%	69%	77%	100%
Dec, 2012	33%	44%	44%	55%	65%	69%	69%	73%	100%

IGNYS's warehouse confirms the incoming orders received from the dealers based on the availability of the material in the warehouse. However state government permit availability is a major issue here because every spare part order requires an official permit of the state without which the material cannot be transported to that state. The permit has to be sent by the dealer, and sometimes the dealer places a lot of orders but does not send the required number of permits to IGNYS's warehouse. In this case, the warehouse cannot make invoices for the orders that do not have permits, and the orders would be delayed for dispatch until the permits are received. Some dealers do not require permits if the orders are placed within the state where IGNYS's warehouse is located. For instance, the orders coming from the dealers in Maharashtra state do not need permits.

Vendors

Vendors are the backbone of the supply chain. They supply parts to Original Equipment Manufacturers (OEMs). In IGNYS's supply chain, vendors supply the main sourcing division and the spare parts division separately, interacting with the individual sourcing managers. For the spare parts division, the vendors take the orders from its sourcing manager and supply them to the mother warehouse. However, the orders given by the sourcing manager fluctuate highly due to demand uncertainty. After discussing with the senior management of the vendors about their problems, the consultant has identified *high back order* as a major problem. It is also a major concern for the senior management of IGNYS. The consultant has also realized that the spare parts sourcing manager does not share dealers' actual ordering data with vendors; hence, the vendors are not able to plan their inventory much ahead of time. The vendors claim that forecasting of spare parts done by the sourcing manager is wrong and erroneous. The vendors also encounter space constraints at the mother warehouse of IGNYS, because unloading the spare parts there usually takes 3–4 days.

On the other hand, IGNYS usually has smooth demand for final goods. Hence, the vendors get almost constant, regular, and confirmed business orders from the main sourcing division, so they give more preference to those orders over the orders from the spare parts division. The logic behind this, according to the vendors, is that the

orders from the spare parts division are irregular; hence, the vendors
cannot achieve proper economies of scale. A select few vendors are
contributing most of the backorder problems. It has been seen that
70–80% of the backorders (by value) come from only 20% of the top
SKUs. Finally, each souring manager of the spare parts division looks
after 30–40 vendors, and sometimes the vendors complain that the
individual attention of the sourcing manager is very poor.

Warehousing

IGNYS has three central mother warehouses at one location for
the spare parts division. The operations are managed by a third party.
All orders come to the mother warehouse from the dealers, and after
the software runs, the spare parts present in the warehouse are automatically assigned to the respective orders. The next day, an operator
collects the parts from the bins in specified locations in the mother
warehouse and gives the parts to prepackaging department. After prepackaging, the order is given to the dispatch office. The part details
in the order are inspected, and the permit and invoice are attached to
the order before dispatching.

After discussing with managers of the warehouses and analyzing
the process flow diagrams at the three warehouses, the consultant has
identified the first major constraint in the warehouse as the *space*.
Also, the warehousing operations are handled by third party; hence,
sometimes there is a conflict of interest with respect to doing work
that is more beneficial to the third party as compared to IGNYS.
Sometimes the spare parts sourcing manager tries to persuade warehousing operators to work on their orders, leaving the operator's job
to arrange and execute the orders on first-come, first-served basis.
Warehouse operators still rely on manual work as the level of automation in the warehouse is low. There is idle inventory, which in some
cases has been lying for many years in the warehouses. Because of
presence of these constraints, too much time is lost in finding the
required parts and filling the orders.

Presently, IGNYS maintains a safety stock of all the spare parts.
The trucks of vendors usually take 3–4 days for unloading the materials at the warehouse, and the vendors are upset because of this time
delay. The spare parts are checked at the main gate, and then a location is assigned to the parts. Next, the logistics person stores the spare

parts at the required destination inside the warehouse. When searching for material, the SAP system provides the location of the spare parts. After the parts are retrieved, they are sent to the prepackaging and post-packaging process areas. These processes include individual material packing, final clubbing according to the orders, attaching permits and printing the invoices. Finally, the orders are dispatched based on minimum ordering quantity.

The Sourcing Division

The main responsibility of the spare parts sourcing division is to take orders from the dealers and place the orders to either the mother warehouse, if material is present at the warehouse, or place the orders with the vendors. The concerns for this division include the high bargaining power of the vendors and late order deliveries. Few vendors have capacity or monitory constraints; hence, they give more preference to the orders received from the main sourcing division over the spare parts sourcing division. The vendors also frequently ask for price increments. The spare parts sourcing division is also concerned about the demand variation and is not able to predict the future demand for spare parts well. Thus, it has become difficult to meet the timelines without having high inventory levels for all the spare parts. Each spare parts sourcing manager handles up to 40 vendors, making the job very difficult. The consultant believes that is the cause of the high attrition rate in the division. The average age in the division is 35, which means that there are many junior people working in the division, and many times they are not able to handle the vendors' concerns. The order forecasting accuracy is very poor, even though the sourcing manager accounts for trend and seasonality. There is also concern about data and knowledge sharing between the vendors and the sourcing manager. Experienced personnel are also reluctant to share data and knowledge with junior members.

Summary

The number of urgent orders has been continually increasing, which is leading to the low service levels in warehousing. The point strongly raised by Ms. Arunima Pandey is that the warehousing operations have gone bad. On the other side, the dealers are unhappy since they are losing their margins on the relax orders and are forced to place urgent orders so that the material arrives on time. The vendors are not happy as they are not getting confirmed and stable orders from the spare parts division. There is a high attrition rate in the sourcing division. The operations at the warehouse are being managed by a low level of automation. The team is now ready for the senior management presentation.

Discussion Questions

1. Identify the undesired effects (UDEs) in the spare parts supply chain of IGNYS. (*Hint*: UDEs are the indicators of problems, have negative influence on a system or a subsystem performance and are not desired on the way of achieving the goal.)

2. Identify the core undesired effect(s) in the IGNYS's spare parts supply chain. (*Hint*: Core undesired effect(s) cause most of the UDEs. Here, to identify the core cause, develop a current reality tree.)

3. Identify the conflicts that can be observed in the supply chain of IGNYS. (*Hint*: Conflicts could be alternative or opposite actions or issues having a common goal.)

4. Suggest the solution(s)/action(s) to solve the conflict(s) that you have observed in the IGNYS's supply chain. (*Hint*: Develop a conflict resolution diagram.)

5. What would be the negative consequences of the proposed solution(s) or action(s) if implemented? (*Hint*: The negative consequences are again UDEs that arise if the solution is implemented.)

6. If the players in the supply chain do not agree that proposed solution will solve the problem, how will you convince them to implement the solution? (*Hint*: Develop a future reality tree (FRT) and apply negative branch resolution (NBR).)

7. Identify the obstacles while implementing the proposed solution(s). (*Hint*: Use the first part of developing the Prerequisite Tree-PRT.)

8. Explain the intermediate objectives and actions to overcome the obstacles that are identified in Question 7. (*Hint*: Use the second part of the Prerequisite Tree-PRT.)

9. Develop a concrete map considering the intermediate objectives, actions, and the need for the actions. (*Hint*: Develop the Transition Tree –TRT.)

10. How could you achieve ongoing improvement in IGNYS's supply chain? (*Hint*: Apply the five steps of the Theory of Constraints.)

Case 29

When a Western 3PL Meets an Asian 3PL, Something Magical Happens

Shong-Iee Ivan Su[†]

Industry Background

In the logistics industry, traditional logistics companies usually have a functional service focus, such as transportation, warehousing, etc., and they gradually extend to more functional service areas due to the needs of customers or market forces. Those firms that have successfully met their customers' and markets' constantly changing needs have developed much better customization and innovation capabilities than their counterparts who could not make the transition. Traditional logistics firms are often called third-party logistics service providers, or *3PLs*. The logistics firms who make the successful transformation to multiple functional service providers usually perform operationally and financially much better than those who cannot switch gears. These firms can be called *advanced 3PLs* (Hertz and Alfredsson, 2003; Su 2011).

In the 1990s and 2000s, manufacturing and service outsourcing has been growing dramatically for western industrial firms and has caused landscape changes in many product supply chains, moving from a regional scope to a global scale. Many U.S. corporations need to face the problem of managing a much longer and more complex global supply chain. This is an area that many firms are not familiar with but now have to deal with on a daily basis.

[†] SCLab, Department of Business Administration, Soochow University, Taipei, Taiwan; sisu@scu.edu.tw

However, many companies do not even handle logistics themselves domestically, not to mention handling much more complex international logistics operations. They constantly ask their current logistics service firms for helps or for solutions. Therefore, more and more US 3PL firms are developing international logistics capability to meet the highly demanding new market needs. They cannot fully expand at once; rather, they need to develop gradually either by setting up their own foreign subsidiaries, or through local acquisitions, or partnering with carefully selected foreign logistics firms.

The logic for developing their own subsidiaries is to seek greater control and a long-term stance in foreign markets. However, this strategy requires higher investment and organizational engagement. If a business does not develop to certain economic scale or as expected, there is a high risk of business failure. On the other hand, the reason for partnering with a foreign logistics firm is to reduce business failure risk to a minimum while still serving the foreign markets without initial heavy capital and organizational investments. However, the key success factor for this strategy is finding a capable and trustworthy foreign partner in the local market.

Several highly recognized US 3PLs have replaced their CEOs with veterans having international logistics expertise and experiences in other regions, such as Asia. It is obvious that these firms are eager to make the right organizational changes with the best top talents on board to lead the transition. Jacobson Companies, for example, is a large US 3PL with a long-term warehousing focus. The board sensed the market needs and in March 2009, hired a new President and CEO with broad-based international logistics experience (Jacobson 2009). More recently, in June 2011, Weber Logistics, also a large warehousing and distribution-focused 3PL, named a new President and CEO who used to work for many years at a large European global logistics firm (AS 2011).

In the following, a 3PL in North America (based in the US) and a 3PL in Asia (based in Taiwan) have been working with each other to better serve the needs of North American customers regarding the logistics and shipping needs between North America and other parts of the world in their supply chains.

Introduction of the Case Companies

Johanson Transportation Service (Justified Timely Solutions)

Founded in 1971 by Richard Johanson, JTS (Johanson Transportation Service) has grown from a local transportation brokerage company to an advanced family-owned third-party logistics service provider helping client companies manage supply chains with "Justified Timely Solutions" that exceed their unique business challenges. Providing unmatched service at a fair price, JTS offers customized freight solutions, including dry and temperature-controlled truckload, less-than-truckload, ocean freight and air; rail/intermodal; and comprehensive importing/exporting solutions with one point of contact. JTS adds value with logistics management, consulting and state-of-the-art technology systems with real-time online tools to facilitate seamless supply-chain communications for its customers. JTS is headquartered in Fresno, CA, and has six regional offices in North America: Sacramento, CA; Tigard, OR; Salem, OR; Madison, WI; Denville, NJ; and, Orange Park, FL (Johanson 2011).

Dimerco Express Group

Established in 1971, DEG (Dimerco Express Group) has been a proactive player in the evolving world of international transportation and logistics. Over the course of 40 years, DEG has progressively expanded its service network on a global scale, especially in Greater China. DEG is publicly listed in the Taiwan Stock market and positions itself as "Your China Logistics Specialist." It is continually expanding marketing and service outlets in strategic locations. DEG's intimate understanding of the ever-changing Chinese logistics market enables its service system to respond expeditiously to clients' requests for effective supply chain management solutions. More recently, the East Asian and Southeast Asian markets are undergoing a major consolidation due to the integration of major free trade agreements and new free trade initiatives in the regions. DEG is expanding aggressively into the Southeast Asian market to provide comprehensive support to customers' growth strategies in Asia.

At the same time, DEG continues to invest and upgrade its e-commerce platform, Dimerco Value Plus System, to meet the needs of global customers. The new system can help customers bolster their competitiveness by providing external data integration and real-time information visibility for effective supply chain management. A better consolidation of the internal data and information also helps to enhance DEG's worldwide management performance (Dimerco 2011).

Search for an International Partner

Since 2000, more and more customers have inquired about international logistics service at JTS. Its management knew that it was the time to seriously consider the development of its international business.

Under the urging of a third-generation family member, a feasibility study on JTS international logistics services was conducted in early 2006 with the support of the Craig Business School of California State University at Fresno. The feasibility study had given a strong recommendation to develop an international logistics capability through an overseas partnership. JTS has since established an international division for international business development and acquired a freight forwarder and NVOCC licenses. Through an extensive search, DEG, an advanced Asian-based 3PL, emerged and was sought after in mid-2006 as an international partner to develop the business cooperation that can complement the strengths of both firms (Negueloua 2011).

Joint Business Development and Complementary Service Offering

Since late 2006, DEG has played the silent role as JTS's international logistics partner to support the ocean and air shipments with the United States as the origin or destination; that is, businesses secured by the JTS International Division. JTS and DEG (SFO branch) often make business calls together and assist each other in bids for both domestic and international logistics demands (Negueloua 2011).

Because JTS and DEG are all advanced 3PL firms, when joined together, they can complement each other and integrate their service networks and service offerings in their focused regions, providing the clients a more comprehensive international logistics service options such as the following (Negueloua 2011):

- Ocean Freight Import: FCL (full container load) & LCL (less than container load)
- Ocean Freight Export: FCL & LCL including Reefer, D.G. Cargo & Project Cargo
- Air Freight Export including D.G. Cargo (dangerous cargo)
- Air Freight Import

Joint Project Illustrations

JTS Client 1—Serving over Four Years

Client 1 needed to deal with growing overseas operations. However, this client had encountered communication challenges in its global supply chain pipeline. JTS has worked with DEG's overseas offices to help the client solve its communication and visibility problems and build a stronger relationship with the customer. The business has been stabilized since then with one to two ocean containers shipping monthly from China to the U.S.

JTS Client 2—New Client

Client 2 needed to use ocean shipping to import oversized/expensive machinery from Europe to the U.S. This client was worried about whether the shipment would arrive on time due to its difficulty in international transportation. JTS communicated closely with the client's requirements on the one hand and used the DEG international logistics network and knowledge to accomplish this project on the other hand. The client had complete visibility of the shipment throughout the whole process, and the shipment even arrived at destination in advance. DEG handled the shipping from the start to the end on behalf of JTS. After the equipment had arrived at the U.S. East coast, it was transferred to a warehouse. DEG contracted a

trucking firm using a flatbed trailer to ship the equipment directly to
the destination in only two days, rather than the traditional rail shipping arrangement that costs less but has a less controllable time to
destination and a higher handling damage risk. The superior service
performance gained high recognition along with gift bags and a thank-
you letter from the client.

JTS Client 3—Lost and Resecured

Client 3 is a wholesale firm in a low price/low margin market.
It used to be a JTS customer but was lost to a large competitor due
to lower bidding price. However, due to the service issue on international shipments, JTS was able to resecure the customer several
years later through the joint network and services with DEG. The
business represents a regular monthly volume of 15 ocean containers
from China to the US.

Mutual Benefits from International Logistics Partnership

Without substantial mutual benefits, it is hard to make two 3PLs
geographically distant from each other dance together harmoniously
in an alliance setup. In the JTS and DEG partnership case, strong
mutual benefits are observed and summarized here:

- More customers' needs are satisfied.
- Customer and market coverage are expanded through a virtu-
ally integrated international logistics network.
- Business philosophies and best practices of the partners can be
exchanged and learned from each other in a constructive way.
- Revenues and profits are increased for both partners.

Discussion Questions

1. From the industry development history and the case story, what
motivates an advanced 3PL to transform from a functional-
focused service firm to a multiple-service firm, and furthermore expand its service network to other regions in the world?

2. How did JTS and DEG develop their partnership relationships? What are the key success factors that can be drawn from this case story?

3. What lessons can be learned from this case? Discuss from JTS's perspective, DEG's perspective, and your own perspective.

References

AS (2011), http://www.americanshipper.com/news/news_intl_story.asp?news=195011.

Dimerco (2011) http://www.dimerco.com/dimerco/en/value&vision.asp.

Hertz S. and Alfredsson M. (2003), "Strategic development of third party logistics providers" *Industrial Marketing Management* 32 (2003) 139–149.

Jacobson (2009), http://www.jacobsonco.com/AboutJacobson/PDFs/CEOPressRelease312.pdf.

Johanson (2011) http://www.johansontrans.com/aboutUs.shtml.

Negueloua, D. (2011), a PPT presentation file on Johanson Transportation Service's Partnership with Dimerco Express Group, 2011 Dimerco Annual Alliance Forum.

Su, S.I. (2011) "An Overview of Third-Party Logistics Industry in a U.S. Context," Working Paper, SC*Lab*, Department of Business Administration, Soochow University.

Case 30

Supply Chain Risk Management for Macro Risks

Matthias Klumpp[†] and Hella Abidi[‡]

Introduction

In recent years, international competition has pressured European companies to improve quality and reduce the time and costs of product development and manufacturing. Furthermore, the economic and financial crisis of 2009 led to an increased effort to outsource manufacturing activities and to find suppliers who can ensure production of products with high quality at lower costs (Kumar 2009). European, in particular German, companies outsource their activities to Asian and other Eastern European countries that provide inexpensive but skilled labor and offer enormous cost reductions. Furthermore, rapid technology development, contracting out, global markets, product dynamics, service complexity, reduced supplier bases, and modern inventory practices are all aspects behind the commonplace complex and interlinked business environment (Deleris and Erhun 2005; Glickman and White 2006).

The corporate strategy of utilizing a global supply chain is afflicted with risks such as linguistic and cultural deficits and customs regulations (Cho and Kang 2001; Schniederjans and Zuckweiler 2004), transportation delays, and logistics service differences (Cho and

[†] Institute for Logistics and Service Management (ild) at FOM University of Applied Sciences, Essen, Germany; matthias.klumpp@fom-ild.de

[‡] Institute for Logistics and Service Management (ild) at FOM University of Applied Sciences, Essen, Germany; hella.abidi@fom-ild.de

Kang 2001). All these types of risks have countermeasures that may increase costs but are typically effective in reducing such risks. On the other hand, *macro risk* factors including (1) natural disasters (floods, earthquakes, hurricanes, fires, and tornadoes), (2) man-made disasters (war and economic crisis), or (3) technical disasters (transport accidents, explosions, fire, gas leaks, and industrial accidents—see Exhibit 30-1) are significant and random events that are especially challenging to manage.

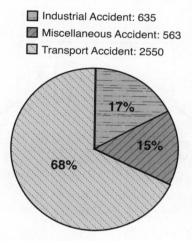

Industrial Accident: 635
Miscellaneous Accident: 563
Transport Accident: 2550

Exhibit 30-1 Number of disasters worldwide.
Source: Figure generated from data found at EM-Dat (2012)

Dramatic collapse of a supply chain due to macro risks argues to verify the strategic, tactical and operational levels of a supply chain and to address all efforts to manage in an efficient way. The three levels in detail are as follows (Kumar 2009):

- *Strategic level:* Is the supply chain aligned with the risk management objectives?

- *Tactical level:* Are all potential risks due to macro risk events well known? Do the supply chain managers have contingency plans in place, and are they prepared when these disasters occur?

- *Operational level:* Is the time known when the prepared contingency plan can be deployed? Are the users able to learn from the experience and to improve their responses to future events?

There are various perceptions in the literature on how to execute risk assessment, risk management and risk mitigation in a global supply chain, and these are common. Risk management should be a continuous and developing process that runs throughout the organization's strategy and the implementation of that strategy (see Exhibit 30-2).

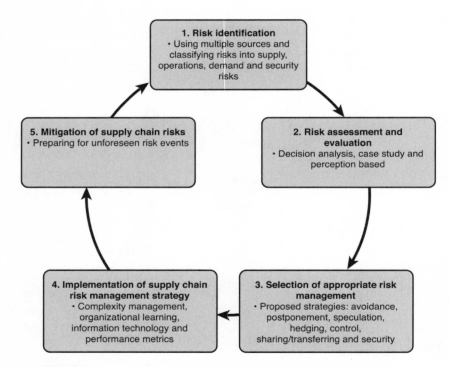

Exhibit 30-2 Five-step process for global supply chain risk management and mitigation

This material is reproduced with the permission of John Wiley & Sons, Inc.

Originally appeared in Manuj and Mentzer (2008), p. 137.

According to IRM/AIRMIC/ALARM (2002): "It [risk management] should address methodically all risks surrounding the organization's activities past, present and in particular, future. It must be integrated into the culture of the organization with an effective policy and a program, led by the most senior management. It must translate the strategy into tactical and operational objectives, assigning responsibility throughout the organization with each manager and employee responsible for the management of risk as part of their job description. It supports accountability, performance measurement, and reward, thus promoting operational efficiency at all levels."

Classification of Risks

Risk mitigation and control has to be classified in two groups: proactive and reactive, because macro risks differ and not all cause catastrophic damage. Because of the increasing number of natural, man-made or technical disasters, the global supply chain has to take measurement regarding how it can reduce vulnerability and build supply chain resilience. The risk factors can be classified in risk classes, which are presented in Table 30-1.

Table 30-1 Classification of Macro Risks as Well as Risk Factors

Risk Class 1	Risk Class 2	Risk Class 3
Relevance category: critical—endangered company value	Relevance category: normal—influenced company value	Relevance category: small—influenced company value
Feature: likelihood >50–75%	Feature: likelihood >26–50%	Feature: likelihood >10–25%
Determinants		
Natural, man-made or technical disasters		
Risk Factors		
Economy risk	Cultural and language difference	Cycle time
Liquidity crisis	Currency decrease	Global risk
Liquidity crisis	Customs regulation	Information management
Price increase	Information management	Inventory management
Process change	Inventory management	Liquidity crisis
Quality	Quality	Quality
Single sourcing	Transportation delay	Trade regulation
Supplier capacity	Nonutilization of transport capacity	Transportation delay

Risk Class 1	Risk Class 2	Risk Class 3
Supplier interdependency		Nonutilization of transport capacity
Transportation delay		
Non-utilization of transport capacity		

Every company has to verify the individual company structure, the supply network, the location of supply chain partners around the globe, and if the location of the supply chain partner is in a country that can be affected swiftly by macro risks. These have a high impact and influence company value as well as turnover in a supply chain. For example, in March 2011, Japan was affected by a tsunami and earthquake, and this caused a high degree of damage in many supply chains of the automotive industry. The supply chain network partners had to procure automotive parts from other suppliers or stop production. Such impacts can cause high costs for the company and threaten the labor market, in particular in the automotive industry.

Therefore the supply chain of different industries has to recognize that there are specific risk determinants where the supply chains have to establish an adapted contingency plan for macro risks. Classification of risk factors to macro risk events by showing the impact on the supply chain can be seen as a result and is helpful for strategic decision making by organizing and issuing a supply chain risk plan for building supply chain resilience.

General Supply Chain Resilience Model

Exhibit 30-3 presents a detailed conceptualization of a supply chain resilience model that considers macro risks. Ways to take action are case-dependent. For example, with a slow onset disaster such as a drought, management has more time to take action and to analyze their location, supplier relationships, contracts, processes, inventory, and demand management than when an industrial or transport accident occurs—in which case they have to act fast and hopefully have a more detailed contingency plan in place. Furthermore, not all risk factors shown in Table 1 are essential and concern macro risks. In summary, slow onset disasters require proactive supply chain risk management, while sudden onset disasters require more reactive supply chain risk management due to missing resources such as time, employees, and money.

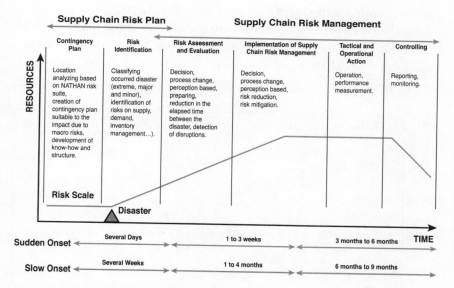

Exhibit 30-3   Supply chain risk management process.

European Case Study

The key actor is a global logistics service provider based in Europe with a turnover of 19.8 billion Euro and 95,000 employees worldwide in 2011. For many years, the logistics provider has cooperated with a Swiss company that is active within the electrical and power plant industry. As one type of cooperation, the logistics provider has introduced a daily transport round-trip from Italy to Germany. Based on a rate agreement, a 20-ton trailer picks up goods from two warehouses of the Swiss company near Rome (Italy) for further European-wide distribution. The transported goods from the two locations near Rome have to be trans-shipped at the logistics provider's terminal in Milano (Italy) for forwarding to the terminal in Mannheim, Germany (central hub).

During the nights of February 4 and 5, 2012, a disaster occurred via a very unusual amount of snow falling and bringing down the very light roof construction of the Swiss company's plant at one of the two origin locations near Rome. This caused a dramatic decrease of shipment volume for the logistics service provider. In particular, the truck from the two locations near Rome to Milano lost three quarters of its utilization, and the same volume was missing for the main haul carriage between Milano and Mannheim. Exhibit 30-4 shows the impact of the disaster.

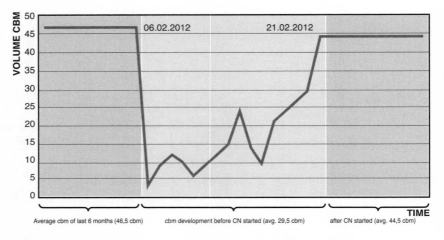

Exhibit 30-4   Volume gap following the disaster (Milano-Mannheim).

On February 21, the Swiss company compensated for the missed volume by importing goods from China for distribution in Europe. But the packages of the goods that arrived from China into the Milano HUB (via Malpensa airport) did not have the same quality compared to ready-to-sell packed goods from the plant near Rome. The logistics service provider had to repackage the imported goods from China to prevent damage and to ensure the promised logistics and transport quality service to the Swiss company.

The disaster caused additional costs. The logistics service provider had to manage the following:

- The volume of shipped goods during the period before the disaster from the two locations near Rome to Milano was 40.2

m^3 and from Milano to Mannheim was 46.5 m^3 on average. From the time until the Chinese imports began, the volumes decreased and reached an average of only 13.7 m^3 on the shuttle and 8.8 m^3 on the prehaulage.

• Within the time of the Chinese imports into the Milano airport, the shipment volume was 44.5 m^3 on average, a little below the normal shipment volume; whereas the volume on the prehaulage was still lower but costs had been covered by the transport costs that were allowed to be charged as per agreed standard rate from the origin plant in order to cover this underutilization and keep the service level steady.

Consequently, the imported goods volume from China did not leverage the gap of the two weeks without any (compensated or uncompensated volume from the collapsed factory). These shortages resulted in the non-utilization of trucks by using the agreed land tariff based on the contract between the Swiss company and logistics service provider. The missing goods volume compared to the average utilization of the shuttle between Milano and Mannheim on a daily basis had a freight rate total amount of 6,811 Euro. The appendix provides a detailed calculation of the determined gap for each individual transport.

Discussion Questions

1. What conclusions can be drawn from the supply chain resilience model (refer to Exhibit 30-3)?

2. What concept is best suited for logistics service providers to reduce the impact of macro risks?

3. What advantages and disadvantages result from the implementation of supply chain risk management systems?

4. For the plant collapse in Italy, can you think of any proactive contingency plans that either the logistics service provider or the Swiss company could have had in place that might have reduced the negative implications from the disaster?

References

Cho, J., Kang, J. (2001): "Benefits and challenges of global sourcing: perception of US Apparel retail firms," *International Marketing Review*, Vol. 18, No. 5, pp. 542–561.

Deleris, L.A., Erhun, F. (2005): "Risk Management in Supply Networks Using Monte-Carlo Simulation," *Proceedings of the 2005 Winter Simulation Conference*.

EM-DAT 2011: The International Disaster Database, retrieved www.emdat.net on 07.10.2012.

Glickmann, T., White, S. (2006): "Security, visibility and resilience: The key to mitigating supply chain vulnerabilities," *International Journal of Logistics Systems and Management*, Vol. 2, No. 2, pp. 107–119.

IRM/AIRMIC/ALARM (2002): A Risk Management Standard, the Institute of Risk Management/Association of Local Authority Risk Managers/Association of Insurance and Risk Managers, London.

Kumar, S.S. (2009): "Risk Management in Supply Chains," *Advances in Management*, Vol. 2, No. 11, pp. 36–39.

Manuj, I., Mentzer, J.T. (2008): "Global Supply Chain Risk Management," *Journal of Business Logistics*, Vol. 29, No. 1, pp. 133–155.

Schniederjans, M.J., Zuckweiler, K.M. (2004): "A quantitative approach to the outsourcing-insourcing decision in an international context," *Management Decision*, Vol. 42, No. 8, pp. 974–986.

Appendix 30-1

Table 30-2 Missing Volumes and Freight Rates Due to Disaster Occurrence

LSP Reference No.	Shipper		Consignee	Colis	Weight	Volume	Tax Weight	Departure Date	Freight MIL-MHG	Freight FTL MIL-MHG	Difference for Compensation
ITMIL002914828	IT	S.	GERMANY	2	628.68	1.536	628.68	2/6/2012			
ITMIL002915285	IT	V.	POLAND	2	28	0.068	28	2/6/2012			
					1231.58	**3.086**	**1244.56**	**2/6/2012**	€ 134.91	€ 1,085.00	€ 950.09
ITMIL002917758	IT	D.	GERMANY	3	93	0.354	106.2	2/7/2012			
ITMIL002918626	IT	V.	UNITED KINGDOM	1	1.6	0.005	1.6	2/7/2012			
					3313.76	**8.489**	**3432.46**	**2/7/2012**	€ 309.68	€ 1,085.00	€ 775.32
ITMIL002920600	IT	S.	GERMANY	13	4086.42	9.984	4086.42	2/8/2012			
ITMIL002921702	IT	V.	DENMARK	1	5.44	0.018	5.44	2/8/2012			
					4844.79	**11.94**	**4870.06**	**2/8/2012**	€ 455.50	€ 1,085.00	€ 629.50
ITMIL002923468	IT	D.	GERMANY	2	71	0.354	106.2	2/9/2012			
ITMIL002924279	IT	V.	POLAND	4	100.5	0.404	121.2	2/9/2012			
					4188.88	**9.411**	**4248.42**	**2/9/2012**	€ 384.36	€ 1,085.00	€ 700.64
ITPDO002926489	IT	M.	BELGIUM	4	286.177	2.669	800.7	2/10/2012			
ITMIL002928558	IT	D.	SWEDEN	2	8	0.053	15.9	2/10/2012			
					1304.84	**5.694**	**1899.2**	**2/10/2012**	€ 203.06	€ 1,085.00	€ 881.94

LSP Reference No.	Shipper		Consignee	Colis	Weight	Volume	Tax Weight	Departure Date	Freight MIL-MHG	Freight FTL MIL-MHG	Difference for Compensation
ITMIL002931466	IT	S.	GERMANY	17	5417.1	7.944	5417.1	2/13/2012			
ITMIL002931534	IT	D.	UNITED KINGDOM	2	42	0.19	57	2/13/2012			
					8007.34	**14.5**	**8054.44**	**2/13/2012**	€ 708.44	€ 1,085.00	€ 376.56
ITPDO002933560	IT	M.	NETHERLANDS	5	997.6	6.64	1992	2/14/2012			
ITMIL002934134	IT	D.	POLAND	1	23	0.095	28.5	2/14/2012			
					6949.84	**23.32**	**8290.74**	**2/14/2012**	€ 746.34	€ 1,085.00	€ 338.66
ITMIL002936684	IT	F.	GERMANY	10	476.4	3.8	950	2/15/2012			
ITMIL002936650	IT	D.	POLAND	1	1.2	0.013	3.9	2/15/2012			
					3732.3	**14.32**	**4472.9**	**2/15/2012**	€ 418.18	€ 1,085.00	€ 666.82
ITMIL002939257	IT	F.	GERMANY	10	532	2.01	532	2/16/2012			
ITMIL002939294	IT	D.	UNITED KINGDOM	3	9	0.066	19.8	2/16/2012			
					3485.07	**9.339**	**3677.39**	**2/16/2012**	€ 341.46	€ 1,085.00	€ 743.54
ITBIT002937253	IT	V.	DENMARK	1	1.14	0.018	5.4	2/17/2012			
ITMIL002942226	IT	D.	POLAND	1	11	0.036	11	2/17/2012			
					4209.7	**20.92**	**5806.28**	**2/17/2012**	€ 533.93	€ 1,085.00	€ 551.07
ITMIL002946364	IT	F.	GERMANY	5	194	0.968	242	2/20/2012			
ITMIL002946475	IT	D.	UNITED KINGDOM	3	45.3	0.226	67.8	2/20/2012			
					9030.96	**29.27**	**9972.18**	**2/20/2012**	€ 887.72	€ 1,085.00	€ 197.28
								TOTAL			€ 6,811

Index

J

Jacket
history, 221
other company strategies
applying, 228
evaluating, 226-227
researching, 222-226
overview, 221
Japanese tsunami and Cisco, 93-95
Jefferson Plumbing Supplies, 117-118
Johansson, Kevin, 111
joint business development
(international logistics partnerships),
270-271
JTS (Johanson Transportation Service)
international logistics partnership, 269
joint business development/
complementary service offerings,
270-271
joint project illustrations, 271-272
mutual benefits, 272
partner search, 270

K

Kerneos, Inc.
overview, 124-125
raw materials, procurement, 126
supplier evaluations, 127
supply chain strategy, 125

L

landfilling, 238
leagile supply chains. *See* hybrid
supply chains
lean supply chains, overstretching
Airbus, 184-186
automobile industry, 192-194
Lerner, Sandy, 81
level production strategies, 130
logistics industry, 267-268

M

macro risks (global supply chains)
classifications, 277
European case study, 279-281
management/mitigation,
implementing, 276-277
overview, 274-277
supply chain resilience model,
277-279
types, 275
manufacturing
3-D printing
3-D Systems website, 182
defined, 181
disadvantage, 182
*traditional production methods,
compared, 181*
clustering, 194
Perdue Farms, 28
*food safety/quality control,
31-32*
hatchery, 29-30
processing, 30-31
process reforms, 212-213
Material Requirements Planning.
See **MRP**
materials
forecast errors, 102
planners, 100
postponing order placements,
122-123
procurement
Kerneos, Inc., 126
NSE, 149-151
process reforms, 213
risks, 65
receiving, 151-152
sourcing. *See* sourcing
medical device industry example. *See*
Remingtin Medical Devices
milk run delivery networks, 136-137

lean supply chains
  Airbus, *184-186*
  automobile industry, 192-194
  operational (Molson Coors), 63-67
    forecasting, 64
    *MRP (Material Requirements
      Planning), 65-67*
  other company strategies
    applying, 228
    evaluating, 226-227
    researching, 222-226
  product de-risk, 92
  product development, planning
    demand forecasting, 105-106
    demand forecasting errors, 101
    demand planning, 105
    end-product planners, 99
    material planners, 100
    materials forecast errors, 102
    order proposals, 100
    production planners, 100
  reactive, 89-90
  resiliency, 92
  SBC platelet supply chain, 42
    collection, 43-45
    issuing process, 47
    *outdate rate and number of
      transfusions statistics, 47*
    overview, 40-41
    platelets, 41-42
    problems, 48
    rotation, 45-47
  sources of risk
    cultural differences, 89
    external, 87
    internal, 86
    natural disasters, 87-88
  supplier financial assessments, 91
  supply chain de-risk, 93
  supply with demand, matching
    electronics risks, 86-93
    Toyota China, 75-79

S

Salvation Army. *See* Dallas Salvation
  Army
SBC platelet supply chain, 42
  outdate rate and number of
    transfusions statistics, 47
  overview, 40-41
  platelets
    overview, 41-42
    shelf life, 42
  problems, 48
  processes
    collection, 43-45
    issuing, 47
    rotation, 45-47
  SBC website, 40
selecting suppliers, 127
sensitivity analysis (eCycle Services),
  115-116
Sherman Soda, 156-159
shipping operations (NSE), 153
Shirley, Lieutenant Eliza, 5
single-sourcing, 222
Sirop, 224-225
sources (risk)
  cultural differences, 89
  external, 87
  internal, 86
  natural disasters, 87-88
sourcing
  demand forecasting, 105-106
  ethics (Fair Trade), 166
    coffee focus, 165
    criticisms, 168
    history, 164
    labeling, 164
    *non-Fair Trade coffee prices
      comparison, 165*
    overview, 163-164
    Starbucks policy, 169-170
    supply chain, 166-167
  multiple, 222

FINANCIAL TIMES

In an increasingly competitive world, it is quality
of thinking that gives an edge—an idea that opens new
doors, a technique that solves a problem, or an insight
that simply helps make sense of it all.

We work with leading authors in the various arenas
of business and finance to bring cutting-edge thinking
and best-learning practices to a global market.

It is our goal to create world-class print publications
and electronic products that give readers
knowledge and understanding that can then be
applied, whether studying or at work.

To find out more about our business
products, you can visit us at www.ftpress.com.